TRANSFORMING COMBUSTION RESEARCH THROUGH CYBERINFRASTRUCTURE

Committee on Building Cyberinfrastructure for
Combustion Research

Board on Mathematical Sciences and Their Applications
Division on Engineering and Physical Sciences

Computer Science and Telecommunications Board
Division on Engineering and Physical Sciences

Board on Chemical Sciences and Technology
Division on Earth and Life Studies

NATIONAL RESEARCH COUNCIL
OF THE NATIONAL ACADEMIES

THE NATIONAL ACADEMIES PRESS
Washington, D.C.
www.nap.edu

THE NATIONAL ACADEMIES PRESS   500 Fifth Street, N.W.   Washington, DC 20001

NOTICE: The project that is the subject of this report was approved by the Governing Board of the National Research Council, whose members are drawn from the councils of the National Academy of Sciences, the National Academy of Engineering, and the Institute of Medicine. The members of the committee responsible for the report were chosen for their special competences and with regard for appropriate balance.

This project was supported by the Air Force Office of Scientific Research under Contract Number FA9550-08-1-0447, the National Institute of Standards and Technology under Contract Number 60NANB9D9023, the National Science Foundation under Contract Number CBET-08333591, and the Department of Energy under Contract Number DE-08NT0007000. Any opinions, findings, conclusions, or recommendations expressed in this publication are those of the author(s) and do not necessarily reflect the view of the organizations or agencies that provided support for the project.

International Standard Book Number-13: 978-0-309-16387-3
International Standard Book Number-10: 0-309-16387-0

Additional copies of this report are available from the National Academies Press, 500 Fifth Street, N.W., Lockbox 285, Washington, DC 20055; (800) 624-6242 or (202) 334-3313 (in the Washington metropolitan area); Internet: http://www.nap.edu.

Suggested citation: National Research Council. 2010. Transforming Combustion Research Through Cyberinfrastructure. Washington, D.C.: The National Academies Press.

Copyright 2011 by the National Academy of Sciences. All rights reserved.

Printed in the United States of America

THE NATIONAL ACADEMIES
Advisers to the Nation on Science, Engineering, and Medicine

The **National Academy of Sciences** is a private, nonprofit, self-perpetuating society of distinguished scholars engaged in scientific and engineering research, dedicated to the furtherance of science and technology and to their use for the general welfare. Upon the authority of the charter granted to it by the Congress in 1863, the Academy has a mandate that requires it to advise the federal government on scientific and technical matters. Dr. Ralph J. Cicerone is president of the National Academy of Sciences.

The **National Academy of Engineering** was established in 1964, under the charter of the National Academy of Sciences, as a parallel organization of outstanding engineers. It is autonomous in its administration and in the selection of its members, sharing with the National Academy of Sciences the responsibility for advising the federal government. The National Academy of Engineering also sponsors engineering programs aimed at meeting national needs, encourages education and research, and recognizes the superior achievements of engineers. Dr. Charles M. Vest is president of the National Academy of Engineering.

The **Institute of Medicine** was established in 1970 by the National Academy of Sciences to secure the services of eminent members of appropriate professions in the examination of policy matters pertaining to the health of the public. The Institute acts under the responsibility given to the National Academy of Sciences by its congressional charter to be an adviser to the federal government and, upon its own initiative, to identify issues of medical care, research, and education. Dr. Harvey V. Fineberg is president of the Institute of Medicine.

The **National Research Council** was organized by the National Academy of Sciences in 1916 to associate the broad community of science and technology with the Academy's purposes of furthering knowledge and advising the federal government. Functioning in accordance with general policies determined by the Academy, the Council has become the principal operating agency of both the National Academy of Sciences and the National Academy of Engineering in providing services to the government, the public, and the scientific and engineering communities. The Council is administered jointly by both Academies and the Institute of Medicine. Dr. Ralph J. Cicerone and Dr. Charles M. Vest are chair and vice chair, respectively, of the National Research Council.

www.national-academies.org

COMMITTEE ON BUILDING CYBERINFRASTRUCTURE FOR COMBUSTION RESEARCH

MITCHELL D. SMOOKE, Yale University, *Chair*
JOHN B. BELL, Lawrence Berkeley National Laboratory
JACQUELINE H. CHEN, Sandia National Laboratories
MEREDITH B. COLKET III, United Technologies Research Center
THOMAS H. DUNNING, University of Illinois at Urbana-Champaign
DENNIS GANNON, Microsoft Corporation
WILLIAM H. GREEN, Massachusetts Institute of Technology
CHUNG K. LAW, NAE,[1] Princeton University
MIRON LIVNY, University of Wisconsin-Madison
MARK LUNDSTROM, NAE, Purdue University
C. BRADLEY MOORE, NAS,[2] University of California, Berkeley
CAROLE L. PALMER, University of Illinois at Urbana-Champaign
ARNAUD TROUVÉ, University of Maryland
CHARLES WESTBROOK, Lawrence Livermore National Laboratory

Staff

SCOTT WEIDMAN, Director, Board on Mathematical Sciences and Their Applications
NEAL GLASSMAN, Study Director
BARBARA WRIGHT, Administrative Assistant

---

[1] NAE, National Academy of Engineering.
[2] NAS, National Academy of Sciences.

BOARD ON MATHEMATICAL SCIENCES AND THEIR APPLICATIONS

C. DAVID LEVERMORE, University of Maryland, *Chair*
TANYA STYBLO BEDER, SBCC Group, Inc.
PHILIP A. BERNSTEIN, Microsoft Corporation
PATRICIA FLATLEY BRENNAN, University of Wisconsin-Madison
EMERY N. BROWN, Massachusetts Institute of Technology
GERALD G. BROWN, U.S. Naval Postgraduate School
RICARDO CABALLERO, Massachusetts Institute of Technology
L. ANTHONY COX, JR., Cox Associates
BRENDA L. DIETRICH, IBM T.J. Watson Research Center
SUSAN J. FRIEDLANDER, University of Southern California
PETER WILCOX JONES, NAS, Yale University
KENNETH L. JUDD, The Hoover Institution
CHARLES M. LUCAS, Osprey Point Consulting
JAMES C. McWILLIAMS, NAS, University of California, Los Angeles
VIJAYAN N. NAIR, University of Michigan
CLAUDIA NEUHAUSER, University of Minnesota
J. TINSLEY ODEN, NAE, University of Texas at Austin
DONALD G. SAARI, NAS, University of California, Irvine
J.B. SILVERS, Case Western Reserve University
GEORGE SUGIHARA, University of California, San Diego
KAREN VOGTMANN, Cornell University
BIN YU, University of California, Berkeley

Staff

SCOTT WEIDMAN, Director
NEAL GLASSMAN, Senior Program Officer
BETH DOLAN, Financial Associate
BARBARA WRIGHT, Administrative Assistant

COMPUTER SCIENCE AND TELECOMMUNICATIONS BOARD

ROBERT F. SPROULL, NAE, Sun Microsystems, Inc., *Chair*
PRITHVIRAJ BANERJEE, Hewlett-Packard Company
STEVEN M. BELLOVIN, NAE, Columbia University
SEYMOUR GOODMAN, Georgia Institute of Technology
JOHN E. KELLY III, IBM Research
JON KLEINBERG, NAE, Cornell University
ROBERT KRAUT, Carnegie Mellon University
SUSAN LANDAU, Radcliffe Institute for Advanced Study
DAVID LIDDLE, US Venture Partners
WILLIAM H. PRESS, NAS, University of Texas at Austin
PRABHAKAR RAGHAVAN, NAE, Yahoo! Labs
DAVID E. SHAW, D.E. Shaw Research
ALFRED Z. SPECTOR, NAE, Google, Inc.
JOHN SWAINSON, Silver Lake Partnership
PETER SZOLOVITS, IOM,[3] Massachusetts Institute of Technology
PETER J. WEINBERGER, Google, Inc.
ERNEST J. WILSON, University of Southern California

Staff

JON EISENBERG, Director
RENEE HAWKINS, Financial and Administrative Manager
HERBERT S. LIN, Chief Scientist
LYNETTE I. MILLETT, Senior Program Officer
EMILY ANN MEYER, Program Officer
ENITA A. WILLIAMS, Associate Program Officer
VIRGINIA BACON TALATI, Associate Program Officer
SHENAE BRADLEY, Senior Program Assistant
ERIC WHITAKER, Senior Program Assistant

---

[3]IOM, Institute of Medicine.

BOARD ON CHEMICAL SCIENCES AND TECHNOLOGY

RYAN R. DIRKX, Arkema, Inc., *Co-Chair*
C. DALE POULTER, NAS, University of Utah, *Co-Chair*
ZHENAN BAO, Stanford University
ROBERT G. BERGMAN, University of California, Berkeley
HENRY E. BRYNDZA, E.I. du Pont de Nemours and Company
EMILY A. CARTER, NAS, Princeton University
PABLO G. DEBENEDETTI, Princeton University
CAROL J. HENRY, George Washington University
CHARLES E. KOLB, Aerodyne Research, Inc.
JOSEF MICHL, University of Colorado
MARK A. RATNER, Northwestern University
ROBERT E. ROBERTS, Science and Technology Policy Institute, Institute for Defense Analyses
DARLENE SOLOMON, Agilent Technologies
ERIK J. SORENSEN, Princeton University
WILLIAM C. TROGLER, University of California, San Diego

Staff

DOROTHY ZOLANDZ, Director
KATHRYN HUGHES, Program Officer
TINA MASCIANGIOLI, Senior Program Officer
ERICKA McGOWAN, Program Officer
AMANDA CLINE, Administrative Assistant
SHEENA SIDDIQUI, Research Associate

Preface

In January 2009, the Multi-Agency Coordinating Committee on Combustion Research (MACCCR) requested that the National Research Council (NRC) conduct a study of the structure and use of a cyberinfrastructure (CI) for combustion research. MACCCR is an informal group of program managers within the federal government that coordinates joint initiatives in basic research involving combustion and keeps group members informed of one another's activities. It consists of representatives from the Air Force Research Laboratory, the Army Research Office, the Office of Naval Research, the Strategic Environmental Research and Development Program of the Department of Defense (DOD), the Energy Efficiency and Renewable Energy Program of the Department of Energy (DOE), the National Energy Technology Laboratory of DOE, the DOE Office of Science, the Federal Aviation Administration, the National Aeronautics and Space Administration, the National Science Foundation, and the National Institute of Standards and Technology.

The combustion research community had already developed a strong foundation for the proposed study through a series of three workshops that shared perspectives and some best practices already developed within portions of the community. Those workshops were held in February and April 2006 and March 2007. They played an important role in exploring selected issues related to CI and in building community interest in the topic.

In response to MACCCR's 2009 request, the NRC assembled the Committee on Building Cyberinfrastructure for Combustion Research under the Board on Mathematical Sciences and Their Applications, the Computer Science and Telecommunications Board, and the Board on Chemical Sciences and Technology to carry out this study. This committee was given the following charge:

1. Identify opportunities to improve combustion research through computational infrastructure (CI)[1] and the potential benefits to applications;
2. Identify the necessary CI elements (hardware, data management, algorithms, software, experimental facilities, people, support, etc.) through examination of existing CI in combustion research and education and CI experience in other, analogous fields. Evaluate the accessibility, sustainability, and economic models for various approaches, and identify positive and cautionary experiences;
3. Identify CI that is needed for education in combustion science and engineering and how education in those fields should change to prepare students for CI-enabled endeavors;
4. Identify human, cultural, institutional, and policy challenges and discuss how other fields are addressing them;
5. Estimate the resources (funding, manpower, facilities) needed to provide stable, long-term CI for research in combustion;
6. Recommend a plan for enhanced exploitation of CI for combustion research, taking into account possible leveraging of CI being developed for computational science and engineering more generally.

In order to conduct this study, the Committee on Building Cyberinfrastructure for Combustion Research met four times between March 9, 2009, and January 20, 2010, in Washington, D.C., and in Irvine, California. It was briefed by representatives of cyberinfrastructures for several scientific communities other than the combustion community and reviewed information provided by these speakers and others.

---

[1] For the purposes of this charge, "CI" is used to abbreviate "computational infrastructure." In the remainder of this report, "CI" stands for "cyberinfrastructure."

Acknowledgments

This report has been reviewed in draft form by individuals chosen for their diverse perspectives and technical expertise, in accordance with procedures approved by the National Research Council's Report Review Committee. The purpose of this independent review is to provide candid and critical comments that will assist the institution in making its published report as sound as possible and to ensure that the report meets institutional standards for objectivity, evidence, and responsiveness to the study charge. The review comments and draft manuscript remain confidential to protect the integrity of the deliberative process. We wish to thank the following individuals for their review of this report:

M. Gurhan Andac, University of Southern California,
Christine Borgman, University of California, Los Angeles,
Sayeed Chaudhury, Johns Hopkins University,
Robert Dibble, University of California, Berkeley,
Rudolph Eigenmann, Purdue University, and
Ruth Pordes, Fermilab.

Although the reviewers listed above have provided many constructive comments and suggestions, they were not asked to endorse the conclusions or recommendations nor did they see the final draft of the report before its release. The review of this report was overseen by Phillip Colella, E.O. Lawrence Berkeley National Laboratory. Appointed by the National Research Council, he was responsible for making certain that an independent examination of this report was carried out in accordance with institutional procedures and that all review comments were carefully considered. Responsibility for the final content of this report rests entirely with the authoring committee and the institution.

The committee also acknowledges the valuable contribution of the following individuals, who provided input at the meetings on which this report is based:

Michael Frenklach, University of California, Berkeley,
Jeffrey Grethe, University of California, San Diego,
Thuc Hoang, Department of Energy,
Walter Polansky, Department of Energy,
Edward Seidel, National Science Foundation,
Douglas Talley, Air Force Research Laboratory,
Phillip Westmoreland, National Science Foundation, and
Frank Wüerthwein, University of California, San Diego.

Contents

SUMMARY 1

1 INTRODUCTION 7
   Alternative Energy Sources, 8
   Cyberinfrastructure in Combustion, 11
   Organization of the Report, 12
   References, 13

2 CYBERINFRASTRUCTURE 14
   Defining "Cyberinfrastructure," 14
   Building a Community Cyberinfrastructure as
      Distributed Collaboration, 20
   The Challenges Facing a Combustion Cyberinfrastructure, 21
      The Petascale Frontier and the Exascale Challenge, 22
      Cyberinfrastructure and Digital Research Collections, 23
      Evolution of Data Collections, 24
      Data-Curation Aims and Challenges, 25
      Filling Data Gaps, 29
      Aligning with the Combustion Community, 30
   Measuring Progress, 32
   Expanding Access to the Community, 32
      Science Gateways, 33
      NanoHUB, 34
      Cloud Computing, 37

Scientific Work Flow, 38
References, 39

3   COMBUSTION AND CYBERINFRASTRUCTURE   41
    Overview, 41
    How Individual Researchers Would Benefit from a
        Combustion Cyberinfrastructure, 43
    A Hierarchical Approach to Combustion Modeling, 44
    Combustor Configurations, 45
    Models, Submodels, and Reductions, 49
    Data and Data Flow, 57
        Species-Based Data, 57
        Chemical Databases, 58
        Continuum-Based Data, 59
        Data Flow, 63
        Cyberinfrastructure: A New Mode of Organization for a
            Community-Level Vision in Combustion Research, 65
    References, 66

4   RECOMMENDATIONS   68
    A Cyberinfrastructure to Connect Combustion Research
        Communities, 70
    Organizational Structure of Proposed Cyberinfrastructure, 71
    Implementation Plan, 72
    A Cyberinfrastructure as an Educational Tool, 74
        Changes in Educational Programs, 74
        Educational Components, 75
    Budgetary Issues, 76
    References, 78

APPENDIXES

A   The GRIMech Model   81
B   CHEMKIN Chemical Kinetics Software   83
C   Direct Numerical Simulations   86
D   Chemical Kinetic Reaction Mechanisms   88
E   Committee Meeting Agendas   91
F   Biographies of the Committee Members   95

Summary

Combustion has provided society with most of its energy needs for millennia, from igniting the fires of cave dwellers to propelling the rockets that traveled to the Moon. Even in the face of climate change and the increasing availability of alternative energy sources, fossil fuels will continue to be used for many decades. However, they will likely become more expensive, and pressure to minimize undesired combustion by-products (pollutants) will likely increase. Even in the absence of fossil fuels, alternative "combustion fuels" such as ethanol or biodiesel fuels are likely to be used and burned in combustion engines, and their use requires the study of the same issues (e.g., burn efficiency, levels of emissions) as those involved in the combustion of fossil fuels.

The trends in the continued use of fossil fuels and likely use of alternative combustion fuels call for more rapid development of improved combustion systems. New engines that are based on more predictive understanding of combustion processes must be designed for new fuel streams. A cyberinfrastructure (CI) that facilitates the timely dissemination of research results, experimental and simulated data, and simulation tools throughout the combustion community and extends into the engineering design process is necessary for shortening the time lines for combustion research (CR), development, and engineering. The current pace is rate-limited—by isolation, replication, and the reliance on experimentation, which is inherently slower than computer simulation. Experimentation is also an expensive but necessary process, especially for engineering design; it may give only limited understanding about by-products and scalability, and it may entirely miss the design optimum. A combustion CI will speed up the process of generating and testing designs and predictions preceding full-scale experimentation.

A cyberinfrastructure is an integrated ensemble consisting of software tools, computing and communication capabilities, and specialized personnel who distribute computing, information, and communication technologies to facilitate the sharing of information, data, software, and computing resources across a community. Examples of such an ensemble are the nanoHUB,[1] a science gateway developed and operated by the National Science Foundation (NSF)-funded Network for Computational Nanotechnology; and the Open Science Grid,[2] a national, shared infrastructure of computing resources funded jointly by the Department of Energy (DOE) and NSF.

A range of architectures, access protocols, management structures, and data models must be considered and crafted for the intended user community when a cyberinfrastructure is being designed. These must likewise be tailored for the specialized needs and culture of the combustion community. Throughout the process of establishing a cyberinfrastructure, care must be taken to ensure buy-in and adoption by the target community.

The combustion research field is well positioned to profit significantly from a community-wide CI. The community is an amalgamation of separate subdisciplines, each with its own data, simulation tools, computing resources, conferences, journals, and cultures. In addition, most CR is conducted by small groups, so the community is not drawn together around large facilities or a small set of research problems. Thus, it can be an unnecessarily slow and haphazard process for one group to learn about and then use improved data from another group. Likewise, simulation tools and research results are not necessarily disseminated as easily and quickly as would be optimal. This "friction" in the system might especially impact the engineering design process, which may not have timely access to the best research, data, methodologies, and tools. The capability of a CI for facilitating the effective sharing of information across the boundaries separating groups and subdisciplines within the combustion community has the potential to transform the community. Throughout this report, the study committee identifies new modes of interaction that a combustion CI will enable, new educational tools that it will make available, and improvements in the process of combustion research that it is likely to spur.

---

[1] See hubzero.org. Accessed October 15, 2010.
[2] See opensciencegrid.org. Accessed October 15, 2010.

At the request of the Multi-Agency Coordinating Committee on Combustion Research, the National Research Council appointed the Committee on Building Cyberinfrastructure for Combustion Research under the Board on Mathematical Sciences and Their Applications, the Computer Science and Telecommunications Board, and the Board on Chemical Sciences and Technology to carry out this study. Appendix F presents biographies of the committee members. This committee was given the following charge:

1. Identify opportunities to improve combustion research through computational infrastructure (CI)[3] and the potential benefits to applications;
2. Identify the necessary CI elements (hardware, data management, algorithms, software, experimental facilities, people, support, etc.) through examination of existing CI in combustion research and education and CI experience in other, analogous fields. Evaluate the accessibility, sustainability, and economic models for various approaches, and identify positive and cautionary experiences;
3. Identify CI that is needed for education in combustion science and engineering and how education in those fields should change to prepare students for CI-enabled endeavors;
4. Identify human, cultural, institutional, and policy challenges and discuss how other fields are addressing them;
5. Estimate the resources (funding, manpower, facilities) needed to provide stable, long-term CI for research in combustion;
6. Recommend a plan for enhanced exploitation of CI for combustion research, taking into account possible leveraging of CI being developed for computational science and engineering more generally.

In order to conduct this study, the Committee on Building Cyberinfrastructure for Combustion Research met four times between March 9, 2009, and January 20, 2010, in Washington, D.C., and in Irvine, California. It was briefed by representatives of cyberinfrastructures for several scientific communities other than the combustion community and reviewed information provided by these speakers and others. Appendix E provides the committee meeting agendas.

The committee's six recommendations are presented below and are discussed in some detail in Chapter 4. Recommendations 1 through 3 concern the need for a combustion cyberinfrastructure and the steps necessary to plan it properly. Recommendations 4 and 6 discuss the funding

---

[3]For the purpose of this charge, "CI" stands for "computational infrastrucure." In the remainder of this report, "CI" stands for "cyberinfrastructure."

and requirements for sustainability of the combustion CI, and Recommendation 5 discusses the functioning of the CI in the research and education communities.

Recommendation 1: A unified combustion cyberinfrastructure should be constructed that efficiently and effectively connects with and enables the movement of data and the sharing of software tools among the different research communities contributing to engine and combustion research and development.

The committee envisions that the proposed CI will be widely adopted by the industrial community, the academic research and development (R&D) community, and the educational elements of the community. Since new data and data that have changed (i.e., parameters associated with chemicals and their reactions) are used so widely, it is crucial that the exchange of data and the tools needed to operate on these data be dependable, rapid, and secure.

The requirements for unification, interface between subcommunities, and validation of the submodels used in combustion simulations lead to the second recommendation.

Recommendation 2: A centralized team will be needed to design and construct a unified, efficient combustion cyberinfrastructure in a timely fashion. At least three individual outreach teams should work closely with a central team: one outreach team connecting with the many chemistry-oriented subcommunities providing fuel data, one team connecting with the reacting-flow and turbulent-flame community, and one team ensuring that the cyberinfrastructure meets the needs of the industrial engine and fuels R&D community. These outreach teams will be responsible for interfaces, specialized software tools, and the development of formats and methods to handle different types of input data, and for the promotion of the new CI within their target communities.

The next two recommendations assign responsibility for planning a CI and encourage the CI's adoption by research funding agencies of the federal government. The development and deployment of a CI for combustion are complex and multifaceted endeavors. Planning alone will be a costly project. In addition to decisions about the required hardware, middleware, and models, decisions must be made concerning what historical data will be put into the system, with accompanying costs and requirements for the curation of these data.

Recommendation 3: Because of the many issues involved in the development and deployment of a CI for combustion, experts in several areas—chemistry data, reacting-flow simulations, engine and fuels R&D, software development, CI maintenance, data curation, deployment, outreach, and education—all need to be involved in the planning, design, and construction of the combustion CI.

Recommendation 4: Federal research agencies responsible for funding combustion research should incorporate specific policies regarding the use of the combustion cyberinfrastructure into their progress reports and their grant processes. The incorporation of such policies will provide incentives to the combustion community and related communities for making the transition to the new system for handling and archiving valuable data.

The proposed CI will also serve as an educational tool. The committee identified two overall educational uses for the CI: (1) as a repository for lectures, courses, and other educational resources for use by graduate students and others—a "combustion portal"; and (2) as a method for the dissemination of the most current information, both through traditional means and through online workshops, courses, symposia, and other presentations—an advanced training program.

Recommendation 5: The combustion cyberinfrastructure should be designed to serve the chemistry and education communities as well as the research community, and to integrate these communities with advances in computer science research and education.

Finally, the issue of the funding level for the CI is addressed. In addition to the funding required to develop and maintain the proposed CI, the committee believes that a mechanism for continued funding should be identified and developed as part of the planning process.

Recommendation 6: A fairly large short-term investment is required to achieve the benefits of a unified combustion cyberinfrastructure. Ongoing operations of this CI will require significant continuing funds. A failure to secure a continuing funding stream to maintain the CI will likely lead to the failure of the whole project.

Combustion will certainly be an important part of the nation's energy infrastructure for decades to come, if not longer. With well-recognized national priorities including energy security, geopolitical stability, and environmental sustainability, there are clear benefits to advancing a strongly supported cyberinfrastructure. This report attempts to give the strong justifications that are needed to invest R&D funds in a combustion cyberinfrastructure.

1

Introduction

Combustion has provided society with most of its energy needs for millennia, from igniting the fires of cave dwellers to propelling the rockets that traveled to the Moon. As newly emerging world economies grow, their energy needs are being met by combustion systems, particularly those burning coal and petroleum-based fuels. Today, the world faces a crisis involving the need to make trade-offs between the ever-increasing energy demands of a growing world and the harmful environmental impacts of chemical pollutant emissions and global warming from greenhouse gases, especially carbon dioxide (CO_2). Science and engineering can help resolve this crisis in many ways, which include improving and replacing today's combustion systems—fossil fuels and biofuels and the systems that extract their energy—to make combustion more efficient and/or cleaner. The pace of developing such improvements or new systems is slower than it need be. A community-wide cyberinfrastructure (CI) that addressed the needs of combustion research and development (R&D) for the timely sharing of data and more powerful simulation capabilities would speed up innovation in all aspects of combustion science and engineering. Software tools, computing and communication hardware, and specialized personnel together would form this critical tool that would enable the delivery of the new technologies that can help resolve this global crisis.

Hydrocarbon fuels and combustion have dominated the world's energy picture for centuries. Coal enabled the industrial revolution and continues to be a major energy source; coal-fired power plants are being built at a dramatic rate today in China (U.S. EIA, 2006) and their emissions are a worldwide concern. Petroleum-based liquid fuels have dominated transportation fuels since the late 1800s, but some estimates suggest that only about a 50- to 100-year supply is left (U.S. EIA, 2007b). The availability of these fossil fuels and their geographical distribution around the world have had profound effects on international politics and economics, as well as enormous impacts on lifestyles everywhere by affecting where and how people live and work. Wars have been fought over access to petroleum and other fossil fuel resources, and growing pressures for energy may again lead to serious international conflicts.[1]

Fossil fuels will not last forever, but it is likely that they will continue to be a primary source of energy worldwide for years to come. There is no alternative energy technology that can replace combustion now, either with respect to the costs of supply or to the total capacity to provide for the huge energy needs of modern society. As fossil fuels diminish, they are likely to be at least partially replaced by other combustion fuels such as ethanol and biodiesel fuels. Energy production technologies change slowly, over decades, because of the enormous investments involved. Time is short, as combustion CO_2 emissions continue to grow explosively, potentially increasing the rate of global climate change that is taking place. All of this makes it essential to revolutionize the ways in which the combustion of these conventional fuels and that of new alternative fuels are achieved; incremental advances will not suffice. As new carbon-neutral fuels are developed, there will be a need to understand how they burn and impact combustion-based engineered systems. The Committee on Building Cyberinfrastructure for Combustion Research believes that a community-wide CI, as presented in Chapter 2, would fuel the transformation that will accelerate the progress in combustion research and development. This transformation will be accomplished through the effective exchange of information, data, and software tools among the various subdisciplines and organizations that form the combustion community, as well as through advances in predictive capabilities that leverage state-of-the-art computer simulations and computing power.

ALTERNATIVE ENERGY SOURCES

An emerging "green revolution" in energy provides some optimism for the partial replacement of fossil fuels by renewable energy sources in the coming years. Wind power and solar power are growing rapidly and may eventually become significant sources of energy, and hydroelectric

---

[1]The role of the need for oil as a cause of war is widely documented. See, in particular, Heinberg (2003).

power has long been a significant part of the U.S. energy supply. However, there are limited opportunities to develop new major hydroelectric sources, and wind and solar power technologies require massive capital investment that makes them unlikely to compete on a large scale with fossil fuels for many years without large subsidies, tax, or regulatory advantages. At present, fossil fuels are much more economical than either wind or solar energy on a per kilowatt-hour basis (U.S. EIA, 2007a). Wind and solar power also have the disadvantage of being intermittent and are not necessarily well matched to conventional energy demands, so some sort of energy storage revolution would be necessary for them to become major contributors to meeting to the world's energy needs.

Renewable fuels for combustion systems for both propulsion and power generation represent another alternative, and many available options are being studied. Although existing commercial biofuels made from food are likely to provide only up to about 10 percent of total U.S. fuel requirements (U.S. EIA, 2007a), significantly larger amounts could potentially be produced using waste biomass (Perlak et al., 2005), with much less impact on land use and food prices; a large R&D effort is already under way to develop cost-effective methods for making new biofuels from this unused resource. There is also considerable research being conducted into even more novel methods for producing alternative fuels—for example, from bio-engineered algae (Sheehan et al., 1998). The introduction of any alternative fuels to the market in large quantities is likely to require modifications both to the engines and to the raw biofuels in order to optimize the system, and so significant new combustion R&D efforts would be required.

In the long term, there are many possible energy sources that could eventually displace fossil fuels, and most of these are being studied seriously. They include nuclear and thermonuclear power production. Although attractive, these options are likely to require very lengthy development times. The needs of power and transportation systems, especially aircraft, will very likely require liquid hydrocarbon fuels for years to come, as the power densities required for flight are high and, at present, cannot be satisfied in any other obvious way. Even after the world has consumed all of its naturally occurring liquid hydrocarbon fuels, there will remain a need to produce such fuels to power aircraft, probably by means of the Fischer-Tropsch[2] or the biofuel production process. This

---

[2]The Fischer-Tropsch process is a set of chemical reactions that convert a mixture of carbon monoxide and hydrogen into liquid hydrocarbons. The process, a key component of gas-to-liquids technology, produces a petroleum substitute, typically from coal, natural gas, or biomass, for use as synthetic lubrication oil and as synthetic fuel.

could change eventually, but the committee has not yet seen any proposed alternative that could become competitive with liquid-fuel combustion.

Even if an economical, nonpolluting, non-greenhouse-gas-emitting energy source were identified and had the potential to replace combustion sources, the overall infrastructure costs of replacing the world's power plants, cars and trucks, furnaces, home heating systems, and other energy systems would be astronomical. The timescales for transitioning to such a new infrastructure would require generations. Today, some power plants and industrial facilities that are more than 75 years old are still in operation; energy infrastructure does not share the short life cycles of some technologies. In addition, most possible future energy technologies are relatively immature, and many years of R&D will be needed to make any such solution viable. Research and development of all types of alternative energy sources must be encouraged, but at the same time people must be realistic about the time constants for the replacement of the dominant technology worldwide that powers our lives, our industries, our homes, and our societies in general.

Nonetheless, it would be dangerous to discount the problems associated with large-scale reliance on today's combustion technologies. They are certainly the major source of greenhouse gases, and although combustion is a lot cleaner than in the past, pollutant emissions from combustion systems are still very serious problems. However, given the previous discussion, it is likely that the world will burn fossil fuels for another century or even longer. Liquid petroleum fuels may be gone in about a hundred years, but world resources of coal are likely to remain for a longer time, and the costs of production of these fuels will make them attractive in economic terms for many years.

If one accepts the projection that combustion will remain a dominant energy and power source for world society for another century, there must be a commitment to making combustion perform much more efficiently, more economically, and in a much more benign way than it does today so as to preserve contemporary ways of life while society waits for new technologies to find replacement energy systems. Without modifications, current rates of emissions of greenhouse gases will likely cause environmental effects, the extent of which may not be known for generations, and other toxic emissions may pollute societies in completely unacceptable ways. The committee believes that this is a crucial point in history and that it is essential to leverage state-of-the-art cybertechnology to develop ways to improve the efficiency of combustion while lessening its detrimental emissions.

CYBERINFRASTRUCTURE IN COMBUSTION

Ongoing research in combustion science and engineering is steadily advancing our abilities to produce energy that is cleaner, cheaper, and more environmentally benign. However, as in other areas of science, this research is only leading to incremental improvements in technologies. For example, the present understanding of burn efficiencies and by-products is limited by the challenges of modeling such complex events. Innovations in combustion systems are often explored through the slow and expensive process of experimentation, which may still give only a limited understanding about by-products and scalability. If combustion could be adequately modeled, parameters could be more readily varied, and better efficiencies and reductions in undesirable by-products could be sought. Similarly, explorations of new fuel molecules depend heavily on experiment because it is not yet possible to predict how a novel molecule will perform in a combustion system. Researchers are handicapped because there is no easy way to find data collected by others, although having such a capability would eliminate redundant experiments and calculations while enabling faster propagation of improved parameter values, models, algorithms, and so on.

An extremely effective way of coordinating available resources in search of a technological solution is through a community-wide cyberinfrastructure. Such an approach collects the contributions of all researchers; channels their research toward major goals; focuses their efforts toward the most productive tasks; facilitates the sharing of tools, data, and computing resources; and reduces the waste of time and effort of conflicting and unproductive activities. In the field of combustion, a few examples of such a community-wide activity from the past can show just how valuable such an approach has been. The CHEMKIN family of simulation tools was developed at the Sandia National Laboratories (Kee et al., 1980), and those tools rapidly became a genuine standard that was used by researchers all over the world. Similarly, the GRIMech chemical kinetic reaction model[3] was developed by researchers funded by the Gas Research Institute (GRI), and it too became a standard used throughout the world combustion community. Both projects were eventually terminated owing to inadequate financial support from their sponsors, and while the CHEMKIN simulation tools were taken over by a private company (Reaction Design, 2009), the community kinetic mechanism activity has not been renewed, despite well-intentioned attempts to revive it (Frenklach, 2007). Technology has reached the point at which it is now possible to conceive of a national,

---

[3]See hppt://www.me.berkeley.edu/gri_mech/version30/text30.html. Accessed October 15, 2010.

or even international, CI. This report suggests that the conditions for CI can be promoted and that the establishment of a CI for combustion research represents the best possible way for the combustion community to contribute expeditiously to reducing the problems that society faces in producing energy.

As is discussed in this report, combustion R&D integrates information from a number of distinct disciplines (chemistry and chemical engineering, mechanical and aerospace engineering, and others) across a broad range of scales, from the atomistic to large-scale engineering systems. At the present time, this data-integration process is slow and tedious, and it only works well when there are cooperative relationships among the researchers working on different aspects of the problem. Progress in combustion R&D could be greatly enhanced by improving how data are integrated both within the individual subdisciplines and, perhaps more importantly, among the disciplines.

A cyberinfrastructure would provide the means needed to exchange data across all of the subdisciplines that are required to make progress in combustion: A properly designed CI could not only facilitate the sharing of data within a subdiscipline but also facilitate the transfer of information among different subdisciplines. Marshaling the power of a CI would dramatically speed the flow of information within the combustion community and consequently significantly decrease the time required to develop new combustion systems that can cleanly and efficiently burn existing fuels while assessing the relative merits of different proposed alternative fuels.

ORGANIZATION OF THE REPORT

This report explains the need for a combustion cyberinfrastructure and the relationships between combustion science and computer science needed to implement such a CI; it also develops a strategic view for the development of the CI. Planning for the CI is beyond the scope of this report; however, the report does recommend steps for the planning process and conditions that need to be met for the eventual CI to be successful.

Following this introductory chapter, in Chapter 2 the committee first defines "cyberinfrastructure" and then addresses the architectural and structural decisions that must be made to construct such a research infrastructure. As seen in Chapter 3, research in combustion is characterized by the interaction of experiment and simulation across a wide range of disciplines and a disparate range of length- and timescales. Data produced by complex simulations or detailed experiments have to flow smoothly throughout the community of researchers and engineers and be processed by advanced software tools that utilize a wide spectrum of processing, storage, and communication capabilities. All of this must be deployed, operated, and maintained by experienced personnel as part of a sustained community-wide effort. Other scientific disciplines have constructed cyberinfrastructures that meet their unique needs; Chapter 3 of this report examines those needs for the combustion community. Chapter 4 contains the recommendations that the committee believes should be followed for the successful construction and operation of a combustion cyberinfrastructure.

This report contains six appendixes that provide the following:

- A discussion of the GRIMech model for the combustion of natural gas,
- A discussion of CHEMKIN Chemical Kinetics Software and its historical background,
- A description of the computation approach, direct numerical simulation,
- A discussion of chemical kinetic reaction mechanisms,
- The committee meeting agendas, and
- Biographies of the committee members.

REFERENCES

Frenklach, M. 2007. "Transforming Data into Knowledge—Process Informatics for Combustion Chemistry." *Proceedings of the Combustion Institute,* Vol. 31, pp. 125-140.

Heinberg, R. 2003. *The Party's Over: Oil, War, and the Fate of Industrial Societies.* Gabriola Island, British Columbia, Canada: New Society Publishers.

Kee, R.J., J.A. Miller, and T.H. Jefferson. 1980. *CHEMKIN: A General-Purpose, Problem-Independent, Transportable, FORTRAN Chemical Kinetics Code Package.* Report SAND80-8003. Sandia, Calif.: The SANDIA National Laboratories.

Perlak, R.D., L.L. Wright, A.F. Turhollow, R.L. Graham, B.J. Stokes, and D.C. Erbach. 2005. *Bioenergy and Bioproducts: The Technical Feasibility of Billion Ton Annual Supply.* Oak Ridge, Tenn.: Oak Ridge National Laboratory.

Reaction Design. 2009. *Chemkin MFC-3.5.* San Diego, Calif.

Sheehan, J., T. Dunahay, J. Benemann, and P. Roessler. 1998. *A Look Back at the U.S. Department of Energy's Aquatic Species Program—Biodiesel from Algae.* Golden, Colo.: National Renewable Energy Laboratory.

U.S. EIA (U.S. Energy Information Administration). 2006. "Country Analysis Briefs: China." August. Available at http://www.eia.doe.gov/emeu/cabs/China/Environment.html. Accessed December 8, 2010.

U.S. EIA. 2007a. *Independent Statistics and Analysis.* Originally published in U.S. EIA, *Annual Energy Outlook 2007,* February 2007. Washington D.C.

U.S. EIA. 2007b. *U.S. Crude Oil, Natural Gas, and Natural Gas Liquids Reserves, 2006 Annual Report.* DOE/EIA-0216. November. Washington D.C.

2

Cyberinfrastructure

DEFINING "CYBERINFRASTRUCTURE"

In 2003 the National Science Foundation's (NSF's) Blue-Ribbon Advisory Panel on Cyber Infrastructure issued a report, *Revolutionizing Science and Engineering Through Cyber Infrastructure*, that defined "cyberinfrastructure" as the "infrastructure based upon distributed computer, information and communication technology" (NSF, 2003, p. 5). Over the 7 years since that report was published, experience has taught us a great deal about what it takes to build and operate successful cyberinfrastructure (CI) for scientific research communities. The primary lesson learned is that it is not sufficient to focus on technology. A successful combustion CI will require the following:

- Deep engagement of the leading scientists in the field, who will supply the models, algorithms, software, data, and other tools that are to be shared through the CI and who will exploit those shared resources to advance combustion research and development (R&D);
- A critical mass of information technology and scientific domain professionals, ideally all at a single location, to manage and guide the CI;
- Resources not only for doing calculations but also for implementing and sustaining a detailed and long-range plan to store and curate the product data so that they can be mined for insight by others in the community;

- Strategy, plans, and resources to foster and coordinate the retention and sharing of experimental data, whether virtual or centralized, some of which might be created only through the prodding of the CI managers;
- A serious commitment and involvement by the research community; and
- A sustained funding model that balances investments in advancing high-risk activities with funds needed to operate a dependable infrastructure that provides the community backbone.

This report refers to all of these elements (hardware, software, data, and personnel) as "the cyberinfrastructure."

Building, operating, and maintaining such a community CI requires a coordinated effort that is fully integrated with the research and engineering vision and roadmap of the community. By its very nature, a CI is a multifaceted entity that spans technology and sociology. In fact, the primary value that a CI can provide to the combustion community is to bridge the disparate subcommunities (kineticists, fluid dynamicists, industrial designers, and so on), and so, by definition, it must be broad and encompassing. This bridging will be achieved by the use of the CI, necessitated by its value to individual researchers. A CI can also require large investments. Although one of the main goals of any community CI is to facilitate the sharing of data and information, an effective community CI also requires the sharing of resources and the adoption of common tools and methodologies. *This entire endeavor can be accomplished only through strong leadership, long-term planning and funding, commitment to cultural changes, and methodical execution.*

The operation of a CI for scientific and engineering research requires the use of a business model for getting the operation funded and for allocating the funds, strategic decision making about which technologies should be used and/or adopted, and a strong and effective executive management. In addition to these functions, building a community CI needs one or more skillful "technology evangelists" who can build support for the CI and a community vision. In order for these components to be effected, there must be a model for how the community functions. Chapter 3 sketches this model for community functioning in broad terms, but an early step in creating a CI for combustion would entail a more detailed exposition of the resources that exist and a characterization of the flows of information through the community, from basic researchers all the way to designers of combustion systems.

The leaders of the combustion CI must take responsibility for sparking a cultural transformation, because the CI can succeed only if the community truly embraces this new model for carrying out R&D. At the same time, those leaders must deal with the cultural transformation caused or triggered by the CI. The foundation for a successful community CI is a culture of sharing that is anchored in mutual trust and respect and a focus on goals that are larger than those required just for creating publishable results. The community CI enables and promotes sharing within and across subcommunities, and membership in the CI assumes an acceptance of this culture of sharing and of mutual goals.

As has been demonstrated by other communities, establishing a community CI can have a transformative impact on all involved. However, by the very nature of a transformative force, building a community CI will require the implementation of "painful" decisions. Some groups will have to give up "ownership" of software tools, computing capabilities, data, or other hallmarks of being a special resource—in some cases because of redundancy and in some cases because of quality or openness. In other words, you cannot have a revolution *and* keep everyone happy.

The CI of the combustion community cannot operate in a vacuum. It will have to interact with other CIs and leverage cyber tools and methodologies developed by other communities. National CIs such as the Open Science Grid[1] and the NSF TeraGrid[2] offer services and computing resources, whereas community CIs provide data and tools. The external CIs also offer help in steering the community CI in the rapidly changing (and in many cases confusing) landscape of hardware and software technologies. Learning from the experience of others and understanding the inherent trade-offs of emerging technology can make the difference between success and failure for a community CI. However, the combustion CI might use extant CIs as models, but it would require its own unique structure. The data requirements, resource requirements, and interconnectivities, as discussed in Chapters 3 and 4, differ in substantial ways from those of the CIs built for other science and technology areas. The field of combustion science is extremely diverse in many aspects (areas, methodologies, size, and so on). As a lower bound on size, the Combustion Institute had 2,000 to 3,000 members in 2004, although these numbers represent only active researchers, and most people are actively doing research only for a short portion of their careers. The memberships of other organizations—such as the International Association for Fire Safety Science, the American Institute of Aeronautics and Astronautics, the American Society of Mechanical Engineers, and the chemistry societies of many nations—have many additional active combustion researchers. In addition, many people working in industrial positions developing

---

[1]See www.opensciencegrid.org. Accessed June 21, 2010.
[2]See www.teragrid.org. Accessed June 21, 2010.

fuel, power, or engine research can be considered within the combustion community.

To meet its diverse challenges, a community CI requires a wide spectrum of innovative capabilities and technology that have emerged in the past decade, including the following:

- Advances in high-performance computing (discussed in the subsection below entitled "The Petascale Frontier and the Exascale Challenge");
- The emergence of data and their analysis as a fourth paradigm of computing (Hey et al., 2009);
- Information and computation grids that link distributed research centers into a single fabric (discussed in the next section, "Building a Community Cyberinfrastucture as Distributed Collaboration"); and
- The use of collaboration and Web technology to create domain data portals for education and outreach (discussed in the subsection below entitled "Science Gateways").

There is a tendency to think that the most important step in building a CI for a community is to acquire a lot of hardware and network it together. However, the actual costs of owning and operating information resources for a research community are in fact dominated by space, cooling, power, and, most important, people. Although federal funding agencies have been willing to provide capital costs for acquiring computing hardware, they often defer to the host institutions to staff, power, and maintain this hardware. This means that many major computational infrastructure projects are funded without a long-term sustainability model to support the most important asset—the human talent that keeps the CI functioning on behalf of the scientific mission. This talent is also responsible for interfacing the CI with the science applications to form effective end-to-end capabilities that leverage the community CI and tailor the CI to support the changing R&D needs. The committee believes that "sustainability" is an attribute of the projected combustion CI that is at least as important as its technical attributes and that sustainability is driven by the CI's human resource needs. At least two prior examples of a CI in the combustion community declined in usefulness—in spite of striking early success—because they were not sustained well (see Chapter 3 in this report). Any plan that is developed must include clear provision for sustainability.

The importance of well-designed and well-tested software infrastructure for building the core of the community CI platform is another major, and often overlooked, category in the total cost of a successful CI. Software that must be used by large research communities and evolve over long time periods must be well designed and reliable. Unfortunately, building reliable software is very hard, and most members of the combustion community are not trained as professional software engineers and may not be rewarded professionally for creating efficient and reliable software. Yet, these scientists are often put in the position of trying to build the software framework needed to support a productive CI. The result is that much software—components that are critical for enabling and accelerating progress in data- and computation-intensive science, such as faster algorithms, data-analysis and visualization tools, middleware, and so on—exists as fragile, "homegrown" products that are difficult to share and which might not work without attention from the person who created the software (that person might have been a graduate student or postdoctoral researcher who has since left). It should be said that such homegrown software does have the potential advantage of being tuned to the researcher's particular needs rather than being more generic. However, that potential advantage does not normally offset the disadvantages. The tendency for homegrown software can also lead to redundant efforts, with the result that the community has multiple versions of software serving a limited number of purposes and none of them being of high quality.

As research collaborations have become large, distributed, and complex, the reliability of the infrastructure has become a critical issue. Without reliability, users lose trust in the system and abandon it; at the least, users must take time that would otherwise be devoted to research in order to reuse someone else's products. When researchers cannot get what they need from the shared infrastructure, they develop their "one-off ad hoc" cyber tools. A CI team that included some dedicated computer scientists and greater use of well-established commercial or open-source platforms (with modest customization) could help address these issues. In addition, the combustion community should agree on a set of community codes to be included in the CI and maintained by CI staff. This difficult issue must be resolved before finalizing the CI architecture, as the acceptance of an existing platform will undoubtedly require compromises in the functionality of the resulting infrastructure, and the trade-offs between these compromises and the ease of use and maintenance of the infrastructure should be clearly understood.

The sharing of software infrastructure, computing resources, and data is often more of a social challenge than a technical one. In the case of software infrastructure, there is a reluctance to use software from another group because there may be no assurance of its sustained support. Also, building a trusting relationship among the different groups involved in contributing to a software stack requires an incremental process that cannot be accomplished in a short time and requires long-term commitment from all parties involved. Federal funding agencies have also been reluctant to fund software maintenance and sustainability. Unless software is an extremely general-purpose product, the critical mass of users needed to attract an open-source community committed to supporting and improving it will not be realized. Commercial vendors will not adopt and support software unless there is a business model sufficiently robust to generate profits. For this reason as well, the CI must have sustainable government support.

Data sharing is a different problem. Because of the publication value of their data, some scientists would rather "share their toothbrush than share their data." However, there is a great deal of community data which, if made available in an easily accessible manner, could accelerate scientific progress. The National Institutes of Health has policies that require some data from funded research be made public.[3] Although this is an enlightened policy, there are no standard mechanisms to provide the metadata[4] to make the data reusable; and building repositories that are robust enough to keep the data alive and available is a substantial CI challenge.

The sharing of processing, storage, and networking resources brings its own unique set of challenges. Such sharing requires policies that are easy to define, implement, and verify in order to control who can use what, how much, and when. Dependable and scalable mechanisms are needed to implement these policies in an environment that crosses the boundaries of administrative domains, technologies, and computing models.

Two aspects of data that are of special importance are their provenance and the control of access. Thus, an underpinning component of any shared infrastructure is identity management, as the "owner" of each request to a data item, software component, or computing resource must be authenticated before the request can be considered for authorization. In the same way that the data and information cannot flow without a network, they also cannot flow without a community-wide authentication system. When intercommunity boundaries need to be crossed, the community authentication system must be capable of supporting the mechanisms needed to facilitate the secure and well-managed mapping of identities. The community also needs to develop processes and procedures to assign memberships and roles to individuals.

Given that investment in a community CI can be significant, it may

---

[3]See www.nsf.gov/bfa/policy/dmp.jsp. Accessed October 15, 2010.
[4]"Metadata" refers to data about data. The term applies mainly to electronically archived data and is used to describe the definition, structure, and administration of data files with all of their contents in context in order to ease the further use of the captured and archived data.

result in a reduction in the funding levels of other activities. Even if this is not the case and the CI funding is "new" money, some community members will consider this money as money that would have been available for their work and is now wasted on CI. It is critical to demonstrate value (return on investment) early. There are many advantages to starting with a small set of engaged customers rather than starting with an overly ambitious "We will build it and they will come" attitude. Winning over the community for a large-scale investment that is labeled as "infrastructure" rather than "high-risk/high-return science" is not an easy task, especially when a significant fraction of the investment (software and people) is not as visible as a big piece of hardware. Unless members of the community experience the value of the CI investment, they will not support the activity. Many scientists and engineers are frustrated with the ever-shifting landscape of cyber technologies, always promising that the next "hot (hardware and/or software) technology" is the "dream technology," while offering very little (if any) help to transition to it.

BUILDING A COMMUNITY CYBERINFRASTRUCTURE AS DISTRIBUTED COLLABORATION

The most important elements of a community CI are the research groups that build the tools and generate the data that support progress in the discipline. A successful CI depends on sustained support for these central groups. But the communities involved in the CIs described below evolved from small communities; they did not start as large ones.

There are many examples of this evolution in the United States, Europe, and Asia. The National Center for Atmospheric Research[5] and Unidata[6] are primary centers for atmospheric research in the United States. These closely aligned organizations are persistent and provide both computation and data resources shared by the whole community. Other large groups, such as the Storm Prediction Center,[7] consist of important specialists who augment the strength of the central facilities; other individual groups work with resources provided by the entire network of resources. As shown in the Figure 2.1, the structure of an overall community CI is a multilevel grid of researchers, specialty centers, and centralized resource providers. The high-energy physics community CI is similar, with central facilities at the European Organization for Nuclear Research[8] and Fermilab.[9] Large Hadron Collider (LHC) data are distributed to a national com-

---

[5] See www.ncar.ucar.edu. Accessed June 21, 2010.
[6] See www.unidata.ucar.edu. Accessed June 21, 2010.
[7] See www.spc.noaa.gov. Accessed June 21, 2010.
[8] See public.web.cern.ch. Accessed June 21, 2010.
[9] See www.fnal.gov. Accessed June 21, 2010.

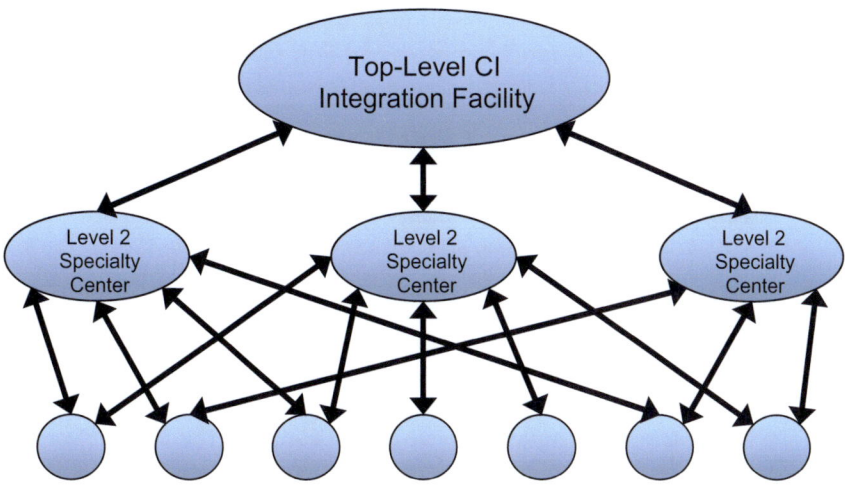

FIGURE 2.1 A typical cyberinfrastructure (CI) organizational chart.

munity of researchers through a multilevel network of data providers. The National Radio Astronomy Observatory[10] provides a similar central organization for the widely distributed radio astronomy community. In each case the technology of CI ties the community CI together.

In Chapter 4, this structure, represented in Figure 2.1, is specialized to the combustion research community.

THE CHALLENGES FACING A COMBUSTION CYBERINFRASTRUCTURE

As the technology of CI has evolved, so has the way that it has been used in science. Until the late 1990s, the use of CI in science was dominated by very large scale numerical simulations done on supercomputers. Innovation was centered on new approaches to parallelizing families of important applications. As the ability to solve more and more complex problems grew, two phase transitions emerged in the way that computational scientists work:

1. *Computation moved into multiscale simulations.* This change often resulted in applications that required a multidisciplinary team of scientists. The applications became more heterogeneous in struc-

---

[10]See www.nrao.edu. Accessed June 21, 2010.

ture, and special middleware[11] was needed to manage them. The teams of scientists working on a problem often lived in different locations, so the CI had to include scientific modes of telepresence and collaboration technology.

2. *The analysis of data became as important as simulation.* An increase in online storage, together with the proliferation of digital instrument streams from sensors in laboratory experiments as well as semi-static data from databases, simulation output, and the World Wide Web, created new CI challenges. New forms of artificial intelligence based on Bayesian statistical methods created revolutionary new ways to discover knowledge from massive data collections in fields such as bioinformatics and natural language translation. The LHC (the largest physics experiment ever devised) is creating the largest data-analysis challenge ever attempted, and the development and maintenance of the associated CI are major components of that project.

The subsections that follow describe the technology challenges that define the frontier of CI in research and education.

The Petascale Frontier and the Exascale Challenge

Combustion-systems modeling is clearly a complex multiscale, multiphysics problem, as described in Chapter 3. Chemical reactions in a combustion device evolve on a nanometer, femtosecond scale, whereas an engine operates at the scale of meters and minutes. It is not feasible to capture this range of scales in a single simulation. Instead, combustion scientists use a range of distinct computational tools along with experimental data, each appropriate to a particular regime, to build models that can be used for simulations at larger scales. Chemical reactions are determined from the synthesis of a broad range of disparate data and simulation tools based on quantum mechanical methodologies such as models for Schrödinger's equation. Fundamental flame properties are determined through a combination of simulations of the Navier-Stokes equations and laboratory flame experiments. Behavior of combustion systems requires complex multiphysics models centered on either Reynolds-averaged or Large-Eddy-Simulations approaches for turbulent computational fluid dynamics (CFD) modeling.

Over the past 20 years, each major increase in computing capabil-

---

[11]Middleware is computer software that connects software components or applications.

ity—from gigaflops to teraflops to, now, petaflops[12]—has led to higher-and-higher-fidelity models of combustion processes, with current models contributing materially to understanding combustion chemistry and physics as well as contributing to the design of more efficient, less polluting combustion devices. However, a great many approximations are still necessary, even on the most advanced supercomputers, and future advances in combustion science will continue to require access to the most powerful computers available as well as to advanced science and engineering codes and applications that can fully exploit the capabilities of these computers.

As new architectures are developed and deployed, a new generation of combustion science software will be required. This next generation of software will need to be more than simply numerical algorithms for solving the core chemistry and physics models needed for combustion. It will also need to encompass new approaches for data analysis and visualization, with an emphasis on making the results of simulations available to the larger community.

In addition, as new architectures evolve, there is a growing consensus that how simulation is approached and what software tools are needed to exploit these architectures fully will have to be rethought at a fundamental level. For exascale computing systems, an international planning activity is currently underway[13] to assess the short-term, medium-term, and long-term software and algorithmic needs of applications for petascale and exascale systems, and to develop a roadmap for software and algorithms on extreme-scale systems. The opportunities provided by petascale and exascale computers for advancing combustion science and engineering are great; however, major investments in software will be required to take full advantage of these new capabilities. In addition, policy issues will need to be addressed, and resources will be required to enable the effective use of these capabilities by academic and industrial researchers.

Cyberinfrastructure and Digital Research Collections

Descriptions of what a CI can do for science are vivid and compelling, with many exemplary cases emerging across domains, as evidenced in a report from the National Science Foundation (NSF, 2008). The advances are profound, but so too are the uncertainties about how to continue to invest in and build a CI in ways that will provide broad support while also enhancing research capabilities for particular research communities

---

[12] "Petascale" refers to computing systems that reach a performance of 10^{15} floating point operations per second. "Exascale" is a 1,000-fold increase over petascale.

[13] See http://www.exascale.org/iesp/Main_Page. Accessed December 10, 2010.

and the work of individual scientists. Supporting and advancing a CI for combustion science need to proceed in step with more global CI development, given the ultimate aim—value-added systems and services that can be widely shared across scientific domains, both supporting and enabling large increases in multi- and interdisciplinary science while reducing the duplication of effort and resources (adapted from the wording in the report of the Blue-Ribbon Advisory Panel on Cyberinfrastructure; see NSF, 2003).

In the general discourse on CIs, the coordination of research data and tools tends to be glossed over as an inevitable trend. But it is often the case that advances in technical capabilities proceed apace while a constellation of underlying social dimensions hinder the growth and use of collections of digital resources. Problems vary across disciplinary cultures, and they are more pronounced in domains that do not have established practices to support the open exchange of publications, software, or data (Kling and McKim, 2000), as is the case with combustion research. Although the present report is concerned with a CI that encompasses more than just data repositories and services, the issues surrounding data are particularly instructive. As seen in the NSF report *Understanding Infrastructure: Dynamics, Tensions, and Design,* many basic CI development challenges are "principally about data: how to get it, how to share it, how to store it, and how to leverage it" for scientific discovery and learning (Edwards et al., 2007, p. 31).

Evolution of Data Collections

The three categories of data collection identified in the report of the National Science Board (2005) entitled *Long-Lived Digital Data Collections: Enabling Research and Education in the 21st Century* provide a framework for considering how digital repositories for combustion research may evolve.

- *Research collections* are the most localized, usually associated with a single investigator or a small laboratory, with limited application of standards, perhaps no intention to archive data over time, and little funding for the management of data. A fundamental aim of the proposed combustion CI is to move beyond this type of distributed, ad hoc approach.
- *Resource collections* are the next step up in coordination, developed for a community of researchers, with standards sometimes developed within the community, but funding sources tend to be variable and unstable.
- *Reference collections* support broad and sometimes diverse com-

munities, conform to well-established standards, and tend to have long-term funding streams from multiple sources.

The Protein Data Bank[14] is the archetype of this last category, reference collections, with its—

- Sophisticated data structures and descriptive vocabularies for protein structures,
- Experimental processes,
- Administrative metadata,
- Highly evolved discovery and access functions, and
- Data validation and security processes (Westbrook et al., 2003; Berman et al., 2003).

As the combustion community works toward reference-level infrastructure, the stages of development will likely follow the path of past successful infrastructures, evolving through the integration of localized systems into a larger functioning network (Edwards et al., 2007). In essence, this course of development brings together (virtually or physically) research-level collections into resource collections to provide the foundation for a more comprehensive reference collection. Table 2.1 captures the relevant features of various structures for data retention and sharing.

Data-Curation Aims and Challenges

Accessible and functional data collections are essential research assets that, if developed through systematic curation principles and methods, will grow not just in size but also in value. Here, "data curation" is defined as the active and ongoing management of data through its life cycle of value to society. This process includes data archiving and digital preservation, but also active appraisal, management, and accountability, provision of context and linkages, and support for reuse and integration (Rusbridge et al., 2005). Data-curation services maintain data integrity and enable data discovery, retrieval, and use over time through a range of activities for identification, selection, authentication, representation, preservation, and other processes that span the entire life cycle of data from production to archiving and reuse. The integration and reuse of data will not happen within a community, however, without a culture of data sharing. At this point in time in combustion research and many other sciences, data sharing is not a commonplace practice or expectation. As is common in many other sciences, data sharing functions more like a cot-

---

[14]See www.pdb.org. Accessed June 21, 2010.

TABLE 2.1 Data Sharing Frameworks and Implications

Aspect	Approach	Characteristics	Implications for Data Producers
Structure	Centralized	Single host location	Deposition services
		Normalized format	Coordinated acquisition location
			Normalized format
	Federated	Single access point	Limited to participants
			Enforced data standards
	Distributed	Individual points of access	Local responsibility for storage
			Control of format
Access	Open	Unlimited access and reuse	Not an option for sensitive data sets
	Hybrid	Access control as needed	Ability to restrict access and use
			Management of sensitive data
	Controlled	Registration and permissions required	Controlled sharing
			Minimized risks
Management	Local	Case-by-case decisions	Control retained
		Potential inconsistency	Maintenance and distribution burden
	Central	Governance by committee or central authority	Policy-driven options for control

SOURCE: Adapted from Pinowar et al. (2008).

Implications for Data Consumers	Considerations for Developers, Service Providers, and Stakeholders
High visibility	Easier coordination of development and maintenance than with other approaches
Optimized retrieval	
	Best for common data types
Enabled browsing	
	Requires sustained funding and personnel
Same attributes as with centralized approach, but with more complex oversight	More complex infrastructure than for centralized structure
	Requires proactive work with participants
	Requires sustained funding and personnel
Low visibility	No control or formal coordination
More difficult retrieval than for other approaches	Rarely maintained for the long term
Complications with interpretation, consistency, integration, and sustainability	
No barriers to participation	Maximizes potential reuse
Some barriers to access	Requires coordination
Access can be complex, time-consuming	Accommodates needs for privacy and security
Ad hoc, inequitable access	Can support gradual transition to more open sharing over time
Guidelines for access and use	Enables consistency
	Potential for building community consensus and standards

tage industry activity, with exchange based on professional relationships and personal communication, and in which actual sharing with anyone other than close, trusted collaborators is negotiated on a case-by-case basis (Cragin et al., 2010).

Much enthusiasm is voiced for the "open data" movement in the flurry of reports from scientific agencies and in the popular scientific press, but studies of data practices point to some of the key obstacles that would be faced in the development of CI capabilities for combustion research. For example, as might be expected, scientists are less inclined to share data that require a laborious data-collection process (Borgman et al., 2007) or which have not been fully mined yet for research results. A recent study across 12 scientific research areas (Cragin et al., 2010) found that willingness to share tends to increase as data are cleaned, processed, refined, and analyzed in the course of research, and particularly after results from data have been published. However, it is not unusual for scientists to withhold data, even within collaborative groups, and to devise strategies to guard against inappropriate use or erroneous interpretation of their data. Moreover, data-sharing enthusiasts can turn into skeptics if they experience some kind of infringement as they begin making their data publicly available. More generally, repositories will not attract contributors unless they have effective access and use policies, with embargo systems to enforce waiting periods before the release of data; also, attribution requirements are of primary importance to many scientists.

Analysis of the roles and functions of nascent public data collections has indicated that criteria-based policies are needed for deposition, sharing, and quality control (Cragin and Shankar, 2006). Policies associated with curation services are complex, as they need to accommodate differences in sharing behaviors and norms, which vary at the subdiscipline level (Research Information Network, 2008), while they also encourage data deposit and use. Since sharing practices are also directly influenced by early experiences, the first steps must be carefully thought out so that early adopters become advocates.

As with all aspects of CI, problems with standards are always looming. There is no consensus on the optimal unit for representing and organizing data, and it can vary within a domain and based on the aims of the depositor or user. This issue is further complicated by the fact that the most presentable or easily analyzed version of a data set may not be the most valuable for preservation over the long term, especially if it is to be integrated or reused by researchers for new types of analysis or in a different discipline (Cragin et al., 2010). Valid interpretation and application by others will require systematic tracking of provenance, determination of the proper targets for archiving and preservation, and consistent application of standards for data capture and for descriptive and contextual metadata. Quality metadata are a linchpin for functioning cyberinfrastructure, but the problems associated with generating metadata have been nearly intractable in many efforts to federate digital resources. As noted by Edwards et al. (2007, p. 31):

> [F]unders have exhorted their grantees to collect and preserve metadata—a prescription that has, for the same number of years, been routinely ignored or under-performed. The metadata conundrum represents a classic mismatch of incentives: while of clear value to the larger community, metadata offers little to nothing to those tasked with producing it and may prove costly and time intensive to boot.

Even in fields where data sharing is accepted practice, the burden of generating contextual metadata to ensure understanding and valid application of data cannot be absorbed by the typical researcher. Scientists rarely have the time or skills needed to prepare data for public sharing (Research Information Network, 2008), resulting in the need for investment in metadata production, preferably at the point of data generation, and a high level of resources directed to supporting functions during the acquisition and ingestion stages of curation (Beagrie et al., 2008).

Filling Data Gaps

A combustion CI also needs to be concerned with filling data gaps, which is not required of some other computational infrastructures referred to above. Models and simulations used for combustion R&D need data as inputs and later for validation, and not all of those data are intrinsically interesting to experimentalists. A full model of combustion might track 1,000 chemical species and require millions of chemical parameters as inputs (see Chapter 3 of this report). Traditional research funding might not be available to provide all of those parameters, and proposals to determine them experimentally might, in many cases, be ranked as low in scientific interest. In other cases, the parameters of real interest for combustion research might be quite difficult to determine experimentally; experimentalists can work with other species to more conveniently examine the underlying science, but the combustion modeler cannot be satisfied with such analogous data.

Thus, for a combustion CI, it should be considered how to provide incentives to fill some gaps in data. Values computed from first principles might someday be feasible, or grants to support "nonglamorous" data collection might be part of the CI plan. Identifying which gaps to fill is in itself a challenging problem.

Aligning with the Combustion Community

To build a CI that supports and advances research, collections need to be aligned with the research process, recognizing that the archiving and access of resources for educational purposes may require different approaches. Nonetheless, shared digital collections can serve a range of functions, all of which have their policy challenges. Some high-value functions for the combustion community include the registration and certification of data sets and an awareness of research production trends, features that have often emerged as unintended outcomes of repository development rather than by design. There is a real opportunity in CI planning to exploit these capabilities to the fullest.

Certain aspects of the field of combustion research make it clearly appropriate as a target for CI development and suggest that a combustion CI has a high probability of benefit and success. These favorable aspects are inherent characteristics of the discipline—not the kind of factors that can be engineered from the outside or imposed by technical structures or policies. The first important factor is the high mutual dependence among subfields. The second is the variation in core problems and methods. Fields with these characteristics are particularly well positioned to capitalize on information exchange systems and likely to have high levels of adoption (Fry, 2006). Moreover, researchers who rely on large bodies of data for modeling and simulation have been early and effective users of public data-repository services. Thus, the combustion community is a relatively low-risk, high-payoff target for a systematic, field-wide approach to collecting, curating, and mobilizing digital collections through a CI.

Known problems in the conceptualization and implementation of distributed, collaborative efforts do pose some risk, however. Areas that need particular attention in supporting combustion research include coordination and cohesion within the community, participation and membership in CI activities, leadership of the initiative over time, the building of trust with all stakeholders, and the translation of the needs of researchers and the technical capabilities of the CI across various disciplinary cultures.

It has been shown that coordination among numerous research sites can hinder collaborative efforts more than the number of disciplines involved, but multiple coordination and communication tactics are effective in improving the management of activities over space and time (Cummings and Kiesler, 2005). Cohesion is essential to building and maintaining effective virtual organizations (NSF, 2008), and it can be strengthened with various techniques, which include making participants' identities explicit and promoting collective activities in the virtual space.

Transparency of leadership, authority, and accountability are necessary and can build trust early on in the development of a community CI. Overcoming distrust after it is established is a much more difficult hurdle. Co-activity across distributed partners is another important strategy that can take the form of training, evaluation, or policy development. Ease of access and participation are key to realizing wide adoption, since more casual interaction (or lurking) is an important way for newcomers to gain entrée to a community of practice (Lave and Wenger, 1991).

Growth and sustainability over time will require a plan for recruiting potential participants with valuable data and software assets for combustion research, as well as researchers in cognate communities who may contribute to or benefit from a combustion CI. Historically, the technologies that provide the foundation for infrastructure have tended to be developed by groups of self-selected experts and enthusiasts for purposes customized for their goals, and whose technological expertise continually increases during the process of development (Edwards et al., 2007, p. 31). The transition from separate, local systems to a pervasive CI will require attracting and supporting the broad range of researchers in combustion and other related scientific disciplines, including more junior scientists and engineers, and those who are more novice users of networked and shared digital resources. Participation can be most readily extended through existing social networks, but attention to incentives for encouraging contributions from more disconnected groups can bring in valuable data, technologies, techniques, resources, and expertise for solving combustion research problems. As linkages and complexity increase, there is a continual need for the translation of requirements and contributions across the fields represented in the growing user base. Translation is also a necessity on the technical side of development, to align data formats, vocabularies, and ontologies for interoperability and integration across domains.

The committee notes that many of the data-curation and -management issues brought up in this section have been anticipated by the NSF through its Sustainable Digital Data Preservation and Access Network Partner (DataNet). Following is a brief description of this project from its prospectus:

> Science and engineering research and education are increasingly digital and increasingly data-intensive. Digital data are not only the output of research but provide input to new hypotheses, enabling new scientific insights and driving innovation. Therein lies one of the major challenges of this scientific generation: how to develop the new methods, management structures and technologies to manage the diversity, size, and complexity of current and future data sets and data streams. This solicitation

addresses that challenge by creating a set of exemplar national and global data research infrastructure organizations (dubbed DataNet Partners) that provide unique opportunities to communities of researchers to advance science and/or engineering research and learning.[15]

MEASURING PROGRESS

Many features and dimensions of a CI can be measured. The subsection below on "NanoHUB" illustrates some of these measures. But benchmarks of real value and impact are difficult and can only be determined over time with good baselines in place. For combustion research, there may be a range of aims particularly worthy of assessment, including the amount of reuse of data and software, new collaborations, citations across subcommunities, the fraction of the research community that uses the CI, and reductions in the amount of time that researchers spend on software and in finding data. The field is still in its infancy, and the means for identifying and measuring such things as results produced with shared data, incentives for participation, quality criteria for data sets, optimal levels of metadata production, and many other issues will need to be resolved in order to optimize the CI to produce the best benefits for the combustion community at large. Sharing of "best practices" with other community CIs would be helpful in this regard.

EXPANDING ACCESS TO THE COMMUNITY

As the combustion community considers the role that cyberinfrastructure might play in its future, certain questions should be asked. On the one hand, are there common data repositories, large-scale instruments, or specialized simulation capabilities that the community as a whole relies on? If so, a compelling case for a shared CI that unites and empowers the community could be made. Or, on the other hand, is the field more characterized by a very wide range of smaller-scale research? If so, a nanoHUB-like model may make sense. Whichever course is taken, a clear understanding of development and support costs should be a starting point. Before launching a costly and extended technology-development effort, one should also examine existing technology platforms that might be leveraged. Below, two of the most prominent platforms that the combustion community might consider as components of its CI are briefly examined: science gateways, represented by the nanoHUB, and cloud computing. The development effort will also require leveraging and link-

---

[15]Available at http://www.nsf.gov/pubs/2007/nsf07601/nsf07601.htm. Accessed on December 6, 2010.

ing to existing facilities such as supercomputing centers and networks of the NSF and the Department of Energy.

Science Gateways

A "science gateway" is a Web-hosted environment for providing users with discipline-specific tools, collaboration, and data. Science gateways usually take the form of a Web portal similar to that of an online bank, an airline, or other retailer. Once the user completes an authentication step ("signs in"), the portal provides the user with his or her personal data and tools to manage the data. In the case of a science gateway, the portal provides access to domain application data as well as tools to manage, generate, and visualize the data. The NSF Tera-Grid project has a very successful ongoing science gateway program. The program has 32 gateway portals, which represent the disciplines of atmospheric science, biochemistry, and molecular structure and function, biophysics, chemistry, earth sciences, astronomy, cosmology, geography, integrative biology and neuroscience, language, cognition and social behavior, neutron science, and materials research. Features common to most gateways include the following: tools for work flow management, so that applications can be rapidly composed from existing components; the ability to personalize the gateway; extensive documentation; and help-desk processes. The TeraGrid organization has a small staff of professionals available to help set up a portal for a science gateway. The staff provides information on best practices for security, job submission to the TeraGrid supercomputers, data management, and Grid protocols. The staff also provides illustrations on how to turn a standard command-line-driven application into a science work flow component.

There are several challenges to making a science gateway a success. Often science gateways are initially funded as an outreach component of a larger grant. They are built by science team members and not by professional developers. Those who build such gateways often start with predetermined concepts of how the software should be built and installed and are not interested in building on the success of others. These gateways seldom integrate well with the TeraGrid security and job-submission protocols. However, with the success of some of the more established gateways, such as nanoHUB and the HUBzero software, more robust portals have emerged. Running a successful portal requires a hosting infrastructure and full-time staff to support it. The greatest challenge comes when the initial funding runs out. TeraGrid has only had limited resources to help with continued support and to keep the portal running; the application community must devote resources, often from other grants, to keep it alive.

NanoHUB

The Network for Computational Nanotechnology (NCN) was launched in 2002 by the National Science Foundation with an objective of accelerating the evolution of nanoscience to nanotechnology and a vision for using what is now called CI to enable new ways for the community to work, learn, and collaborate. A major goal was to lower barriers to the use of simulation in newly emerging fields of research, thereby fostering collaborations between experimentalists and computational experts and promoting the use of live simulations in education. Toward that goal, the NCN created nanoHUB.org, a science gateway where users log on, access research-grade software being developed by their colleagues in nanotechnogy, run interactive simulations, and view the results online, with no need to download, install, support, and maintain software. A powerful open-source software development platform, rappture.org, makes it easy for developers to create and deploy new codes and to equip them with friendly interfaces designed for nonexperts. The underlying technology platform, HUBzero, now powers eight different "hubs" in various scientific and engineering disciplines and is being readied for an open-source release.[16]

At its core, a "hub" is a Web site built with familiar open-source packages—a Linux system running an Apache Web server with a Lightweight Directory Access Protocol[17] for user logins, PHP[18] Web scripting, Joomla![19] content-management system, and a MySQL[20] database for storing content and usage statistics. The HUBzero software builds on that infrastructure to create an environment in which researchers, educators, and students can access tools and share information.

The signature service of a hub is its ability to deliver interactive, graphical simulation tools through an ordinary Web browser. Unlike a portal, the tools in a hub are interactive; users can zoom in on a graph, rotate a molecule, probe isosurfaces of a three-dimensional volume interactively, and so on, without having to wait for a Web page to refresh.

---

[16]See HUBzero.org. Accessed December 10, 2010.

[17]A Lightweight Directory Access Protocol is an Internet protocol that e-mail and other programs use to look up information from a server.

[18]PHP is a widely used general-purpose scripting language that is especially well suited for Web development.

[19]Joomla! is an open-source content-management-system platform for publishing content on the Web.

[20]MySQL is a relational database management system.

Users can visualize results without having to reserve time on a supercomputer or wait for a batch job to engage. Users can deploy new tools without having to rewrite special code for the Web. The HUBzero tool execution and delivery mechanism is based on Virtual Network Computing (VNC).[21] Any tool with a graphical user interface can be installed on the hub and deployed within a few hours. For legacy tools and other codes without a graphical interface, an interface can be quickly created by using HUBzero's associated Rappture[22] toolkit. The Rappture interface helps to set up jobs and visualize results. The jobs themselves can be dispatched to the TeraGrid, the Open Science Grid, and other participating cluster resources. Using this architecture, the nanoHUB has brought more than 100 different simulation tools online in just a few years, with more tools currently under development. In addition to providing for online simulation, the nanoHUB has also become a major platform for disseminating new research methods and new approaches to education in nanotechnology.

Richard Hamming, the founder of the Association for Computing Machinery, famously said: "The purpose of computing is insight—not numbers." Some of the most popular resources on nanoHUB.org are tutorials, short courses, and full-semester courses that convey new ways of treating the new problems and possibilities associated with nanoscience and technology. Online research seminars (also available as podcast seminars), tutorials, and courses attract a high percentage of nanoHUB users. Complementing the online simulations, seminars, and courses is a set of services for supporting users and resource contributors along with services to manage the development and deployment of software and related services. Content ratings and tagging, citations, wikis and blogs, user-managed groups, and usage metric collection and reporting are also part of the HUBzero infrastructure.

More than 154,000 people (60 percent outside the United States) make use of nanoHUB services each year. More than 1,800 science and education resources have been created, and the nanoHUB has been cited in 575 research publications. As of the 2009-2010 academic year, 320 classes at

---

[21] Virtual Network Computing (VNC) is a graphical desktop-sharing system that is used to remotely control another computer. It transmits the keyboard and mouse events from one computer to another, relaying the graphical screen updates back in the other direction, over a network.

[22] The Rappture (Rapid APPlication infrastrucTURE) toolkit provides the basic infrastructure for the development of a large class of scientific applications, allowing scientists to focus on their core algorithm.

130 institutions have used the nanoHUB, including all top 50 U.S. engineering schools.[23]

The nanoHUB is still a work in progress, but some lessons have been learned and some software potentially usable by a broader community has been developed. The nanoHUB science gateway has proven to be an effective platform for knowledge dissemination by experts, for education by self-learners, and for the promotion of collaborations between computational experts and experimentalists. The nanoHUB cannot yet be considered to be an online user facility, and it does not yet address the needs of the serious computational scientist. It does well at the dissemination of new research methods—especially those that are concretely instantiated in a simulation tool that often meets the needs of experimentalists and promotes collaboration with them.

As the field matures, computational demands are increasing, and the nanoHUB is being challenged to make cloud computing work for a community of people focused on solving problems and exploring ideas rather than on computational science per se. Still, the value of small-scale computational tools should not be underestimated. In fact, a recent survey on how scientists use computers concludes:

> [I]f funding agencies, vendors and computer science researchers really want to help working scientists do more science, they should invest more in conventional small-scale computing. Big-budget supercomputing projects and e-science grids are more likely to capture magazine covers, but improvements to mundane desktop applications, and to the ways scientists use them, will have more real impact. (Wilson, 2009, p. 360)

An interesting and unanticipated development has been the use of the nanoHUB as a forum for innovative new approaches to teaching concepts in nanoscience and technology. For instance, the initiative "Electronics from the Bottom Up" at the nanoHUB is an effort to rethink the teaching of nanotechnology beginning at the molecular level and working up to the macroscopic level. The collection of courses in this initiative attracts approximately 25,000 people per year. Other examples in teaching innovation include efforts to cross disciplinary boundaries. For example, the educational materials on electronic biosensors are specifically designed to make use of the concepts that electrical engineering students understand while teaching them the relevant concepts from biology. Such multidisciplinary courses are rare, and the people who can develop them well

---

[23]Personal communication to the committee from Gerhard Klimeck, Director, Network for Computational Nanotechnology, December 2010.

are also rare. Because of this, deploying these kinds of courses online for self-learners can have great impact for a community.

Cloud Computing

Science gateways and portals provide a Web interface to scientific data and data-analysis tools. These tools are traditionally hosted on small servers that sit in a researcher's laboratory. This is a reasonable solution as long as the number of users is small. However, if the gateway becomes popular, this method of supporting it becomes less realistic. Cloud computing uses the resources of a large, industrial-scale data center to provide scalable, reliable access to data storage and computational resources. The commercial data center provider charges on a pay-as-you-go basis. If a person only uses one server and a small database, that person is only charged for that service. However, the data center will automatically add servers and capacity as demand for anyone's Web site grows. The name "cloud computing" comes from the idea that the application provider no longer needs to support or maintain the infrastructure; it is a "cloud" of remote resources. The concept is an evolution of the Web-server-hosting industry combined with the infrastructure that was built for large online business such as Web search. Companies such as Google, Microsoft, Amazon, Yahoo!, e-Bay, and IBM have together deployed millions of servers, and they have realized that they can easily provide this massive infrastructure as a service to aid their customers.

Cloud computing is still very new, and its use in scientific applications is just beginning. The NSF has sponsored the Cluster Exploratory (CLuE) program to encourage researchers to experiment with clouds provided by IBM, Google, and others. More recently the NSF announced the Computing in the Cloud program in collaboration with Microsoft. As part of President Obama's initiative to improve government efficiency, the General Services Administration has just introduced Apps.gov, a federal cloud resource for business applications, productivity tools, information technology services, and social media applications.

Cloud computing is not intended to replace traditional supercomputing. At first glance, the architecture of a modern supercomputer and a cloud data center are similar; however, they are optimized for different uses. Supercomputers are designed to support massively parallel simulations for which processors operate in tight synchronization. Cloud data centers support loosely coupled, massively parallel access to large volumes of data and millions of concurrent users. While these two architectures may converge in the years ahead, there are still many applications of both in modern digital science.

SCIENTIFIC WORK FLOW

As science has become more multidisciplinary and voluminous data of different types are more readily available, the nature of the scientific software application has changed from tasks carried out by monolithic FORTRAN programs to those accomplished by complex collections of preprocessors, applications, postprocessors, databases, and visualization tools. The scientific community has recognized that the orchestration of these tasks, known as work flow management, which often has to be applied to hundreds or thousands of input scenarios, is best accomplished by an automated process. A work flow system is a type of programming tool that was invented in the commercial sector to manage large, complex, business-critical processes. In the sciences, work flow systems are used to automate the process of taking scientific input data and moving it to storage, scheduling it for any required preprocessing so that it may be used as initial conditions for large simulations, running the simulations and postprocessing the results. This is especially important when the same sequence of tasks must be accomplished for hundreds of different input data collections.

In the early 2000s, a number of scientific work flow tools were developed (Taylor et al., 2007). For example, a work flow tool has been used to orchestrate the large-scale data movement, data transformations, and visualization in large, turbulent-combustion computations (Chen et al., 2009). By 2005, there were over a dozen different work flow systems designed for specific disciplines. Many were very similar in function, and few were built based on proven industry standards. Each discipline insisted that its requirements were different, and the tools that emerged were often so specialized that they could not be used in a new context. As of this writing, only a few scientific work flow tools remain, and the sustainability of those is questionable. The private sector now provides work flow systems suitable for scientific applications. In some cases a commercial system may be provided at no cost to researchers. In other cases a few standards-based, open-source tools have a large enough user base to be sustained. But, as with much of the successful open-source software, the programmers maintaining it are provided by companies who have a business interest in seeing it continue.

It is entirely possible that there will emerge a single scientific work flow system with enough flexibility to support most science domain challenges. The funding agencies, working with partners in the private sector, can help make this happen. However, this outcome would take an effort by each scientific community to abandon software developed by that community for a particular purpose and to adopt software, possibly developed by other communities for related purposes.

REFERENCES

Beagrie, N., J. Chruszcz, and B. Lavoie. 2008. *Keeping Research Data Safe: A Cost Model and Guidance for UK Universities*. Bristol, United Kingdom. Joint Information Systems Committee. Available at http://www.jisc.ac.uk/publications/reports/2008/keepingresearchdatasafe.aspx. Accessed December 10, 2010.

Berman, H., K. Henrick, and H. Nakamura. 2003. "Announcing the Worldwide Protein Data Bank." *Nature Structural Biology* 10(12):980.

Borgman, C.L., J.C. Wallis, and N. Enyedy. 2007. "Little Science Confronts the Data Deluge: Habitat Ecology, Embedded Sensor Networks, and Digital Libraries." *International Journal of Digital Libraries* 7(1-2):17-30.

Chen J.H., A. Choudhary, B. de Supinski, M. DeVries, E.R. Hawkes, S. Klasky, W.K. Liao, K.L. Ma, J. Mellor-Crummey, N. Podhorszki, R. Sankaran, S. Shende, and C.S. Yoo. 2009. "Terascale Direct Numerical Simulations of Turbulent Combustion Using S3D." *Computational Science and Discovery*, Vol. 2.

Cragin, M.H., and K. Shankar. 2006. "Scientific Data Collections and Distributed Collective Practice." *Computer Supported Cooperative Work* 15(2/3):185-204.

Cragin, M.H., C.L. Palmer, J.R. Carlson, and M. Witt. 2010. "Data Sharing, Small Science, and Institutional Repositories." *Philosophical Transactions of the Royal Society A*. 368(1926):4023-4038.

Cummings, J.N., and S. Kiesler. 2005. "Collaborative Research Across Disciplinary and Organizational Boundaries." *Social Studies of Science* 35(5):703-722.

Edwards, P.N., S.J. Jackson, G.C. Bowker, and C.P. Knobel. 2007. *Understanding Infrastructure: Dynamics, Tensions, and Design. Report of a Workshop on History and Theory of Infrastructure: Lessons for New Scientific Cyberinfrastructures*. Arlington, Va.: National Science Foundation. Available at http://hdl.handle.net/2027.42/49353. Accessed December 10, 2010.

Fry, J. 2006. "Scholarly Research and Information Practices: A Domain Analytic Approach." *Information Processing and Management* 42(1):299-316.

Hey, T., S. Tansley, and K. Tolle, eds. 2009. *The Fourth Paradigm: Data-Intensive Scientific Discovery*. Redmond Wash.: Microsoft Press.

Kling, R., and G. McKim. 2000. "Not Just a Matter of Time: Field Differences and the Shaping of Electronic Media in Supporting Scientific Communication." *Journal of the American Society for Information Science and Technology* 51(14):1306-1320. Available at http://xxx.lanl.gov/ftp/cs/papers/9909/9909008.pdf. Accessed December 10, 2010.

Lave, J., and E. Wenger. 1991. *Situated Learning: Legitimate Peripheral Participation*. Cambridge: University of Cambridge Press.

National Science Board. 2005. *Long-Lived Digital Data Collections: Enabling Research and Education in the 21st Century*. Available at http://www.nsf.gov/pubs/2005/nsb0540/. Accessed January 18, 2011.

NSF (National Science Foundation). 2003. *Revolutionizing Science and Engineering Through Cyber Infrastructure: Report of the National Science Foundation Blue-Ribbon Advisory Panel on Cyberinfrastructure*. Arlington, Va.: National Science Foundation. Available at http://www.nsf.gov/cise/sci/reports/atkins.pdf. Accessed December 10, 2010.

NSF. 2008. *Beyond Being There: A Blueprint for Advancing the Design, Development, and Evaluation of Virtual Organizations*. Final report from Workshops on Building Virtual Organizations. Arlington, Va.: National Science Foundation.

Pinowar, H.A., M.J. Becich, H. Biolfsky, and R.S. Crowley. 2008. "Toward a Data Sharing Culture: Recommendation for Leadership from Academic Health Centers." *Public Library of Science Medicine* 5(9):e183.

Research Information Network. 2008. *RI News Issue 5*. Available at http://www.rin.ac.uk/resources/print-newsletter/rinews-issue-5-summer-2008. Accessed January 18, 2011.

Rusbridge C., P. Burnhill, and S. Ross. 2005. "The Digital Curation Centre: A Vision for Digital Curation." In *From Local to Global: Data Interoperability—Challenges and Technologies*. Available at http://eprints.erpanet.org/82/01/DCC_Vision.pdf. Accessed December 19, 2010.

Taylor, I., E. Deelman, D. Gannon, and M. Shields, eds. 2007. *Workflows for e-Science*. New York, N.Y.: Springer Press.

Westbrook, J., F. Zukang, L. Chen, Y. Huanwang, and H.M. Berman. 2003. "The Protein Data Bank and Structural Genomics." *Nucleic Acids Research* 31(1):489-491.

Wilson, G. 2009. "How Do Scientists Really Use Computers?" *American Scientist*, Vol. 97, p. 360.

3

Combustion and Cyberinfrastructure

OVERVIEW

Combustion has been a scientific and engineering discipline for hundreds of years. The chemical and physical aspects of the field are well known. Today's central challenges for combustion research are, in principle, easily summarized. Revolutionary changes are needed in the way that fuels are converted into work and heat to significantly reduce carbon dioxide (CO_2) and other emissions into the atmosphere while at the same time significantly increasing combustion efficiency so as to make fuels last longer.

Most combustion systems are based on technologies that are very old; gasoline and diesel engines were invented more than 100 years ago. Although they have been improved and refined continuously, they continue to emit toxic chemicals and greenhouse gases, especially CO_2. Factory burners, boilers, and other industrial combustion systems are based on even older technology and still produce carbon black (soot) along with toxic chemicals and CO_2. To the extent that these systems rely on petroleum—which is the case for most transportation, in particular—they are using up a limited resource that by some estimates will be largely depleted in 50 to 100 years (U.S. EIA, 2007). Thus, it is essential to improve the efficiency with which they use that resource. At the same time, indications are that the world cannot tolerate another hundred years of the toxic emissions and global warming currently produced through petroleum combustion (U.S. Global Change Research Program, 2009). Mere tinkering with extremely mature engine and burner technologies will not solve existing combustion problems. A fundamental change in the way that combustion is studied will be required to meet these societal goals. Unfortunately, combustion is a complex problem involving enormous ranges of temporal and spatial scales and a wide range of disparate disciplines. Historically, these different disciplines have studied different aspects of the combustion problem independently, with relatively loose coupling between disciplines. Traditionally this decoupling has been particularly acute between the communities dealing with the detailed characterization of molecular properties and the large-scale continuum modeling community that is focused on the complexities of modeling turbulent flows. This decoupling has arisen primarily because the tools to bridge the gap between molecular properties and turbulent flows were not available.

Two developments in recent years that have had major impacts on progress in science and engineering have been laser diagnostics and high-performance computing (HPC). Used together, these two technologies have accelerated the pace of research and development (R&D) in many important fields, including combustion research, where they offer the potential to bridge the gap between different disciplines that study different facets of the combustion problem. Laser diagnostics have made it possible to examine combustion systems at the molecular level and in complex physical systems where in situ observations had not been possible. For example, it is now possible to probe turbulent flames experimentally in ways that elucidate turbulent-flame structure in detail both spatially and temporally. In addition, time-resolved velocity fields and two-dimensional planar images of flame markers to capture the interaction of a flame with turbulent flows can now be measured.

Computational combustion is quite well established (Westbrook et al., 2005), but it is only within the past decade that HPC has progressed to the point that detailed simulations of the physical and chemical structure of combustion have become possible. Modeling a wide range of two-dimensional combustion problems has now become relatively routine. It has even become possible recently to model turbulent flows at the laboratory scale with detailed chemical kinetics and transport (Puduppakkam et al., 2009). This knowledge is essential in exploiting the information obtained by sophisticated observations and optical diagnostics to bring together the different communities working on combustion and to help effect the necessary revolution in combustion science.

However, to realize this potential and to achieve significant advances, a much more coherent approach to simulations and data management than exists today is needed. Currently, most combustion investigators work independently or as part of small research groups, each of which develops its own simulation tools and has access primarily to its own data. Every group saves experimental and computational data in its own style and format, and many groups develop their own software tools in their own idiosyncratic manner. For example, almost every chemical kinetics group in the world has developed its own submechanisms for the oxidation of hydrogen, methane, propane, and other fuels, rather than each contributing to a collective kinetics model that can be revised by everyone and used by everyone. The only significant exception occurred when a common kinetic mechanism was developed: the GRIMech (Frenklach, 2007) mechanism for methane and natural gas combustion (see Appendix A in this report), assembled by a team of kinetics experts from multiple research groups and optimized exhaustively using results from many experimental studies. That reaction mechanism was responsible for an enormous number of excellent subsequent studies and provided a thoroughly tested research tool that was used worldwide. GRIMech played a significant role in unifying the community before funding to support continued development of the mechanism was lost.

Another example of a community-wide resource was the CHEMKIN family of combustion software tools developed at the Sandia National Laboratories in Livermore, California (see Appendix B in this report). For many years, CHEMKIN provided a valuable common resource worldwide that was used in countless computational combustion studies. The CHEMKIN software tools were particularly important because a combustion researcher could walk into any combustion facility or office in the world and immediately be capable of carrying out modeling studies of simple zero- and one-dimensional combustion problems, since the software tools were the same everywhere.

HOW INDIVIDUAL RESEARCHERS WOULD BENEFIT FROM A COMBUSTION CYBERINFRASTRUCTURE

As has been found in other scientific and engineering disciplines, the coordination of computer software tools, data storage, formats, and protocols can be extremely effective in increasing dramatically the rate of progress in a given field. A well-designed cyberinfrastructure (CI) applied to combustion is a means of enabling this coordination. Among the features of a cyberinfrastructure that would immediately be very helpful would be common repositories and/or registries for experimental and computational data, which would make data sets available to any researcher. The establishment of collaborative groups working to produce, improve, and maintain these common data sets would lead to more accurate and robust models for chemical kinetics, molecular transport, radiation parameters, and thermochemical databases. Similar collaborative groups working at the continuum level would lead to better models for turbulence and other submodels needed for simulating realistic combustion systems. In addition, a suitably designed CI would facilitate interchange among the different disciplines that would serve as a catalyst to change fundamentally the way that combustion research is pursued. Communication of robust and well-validated software tools and databases from developers to users would also be greatly facilitated, and improvements made by one group or individual would immediately be available to many other researchers. As discussed in Chapter 2, the existence of a community CI facilitates such interaction.

A HIERARCHICAL APPROACH TO COMBUSTION MODELING

The starting point for computer simulations of gas-phase combustion is the compressible Navier-Stokes equations, which describe the conservation of mass, momentum, and energy in a compressible fluid. For combustion applications, this system is augmented with additional chemical species equations that describe the transport and reactions of the chemical components in the gas. To examine even a simple laboratory problem such as a steady laminar Bunsen flame, a detailed description of the properties of the gas under consideration must be developed. The thermodynamic behavior of the gas and the chemical kinetics that describes the burning process must be characterized. Also needed are transport properties that specify the viscosity and thermal conductivity of the fluid as well as interspecies diffusive processes. In addition, radiative heat transfer properties are needed to capture accurately the thermal environment of the flame.

For more realistic combustion systems, it is necessary to introduce additional complexity to the model. Most practical combustion devices operate in a turbulent environment; consequently, even for gaseous combustion, one must deal with the complexities of turbulent flows. Many systems, such as internal combustion engines or jet engines, burn liquid fuels instead of gaseous fuels. Such fuels are introduced into the system as a spray, requiring a treatment of multiphase effects. Other types of systems use particulate fuels, such as pulverized coal in a coal-fired burner or solid fuels in a rocket. The formation of soot during the combustion process introduces additional multiphase effects.

This complexity means that a wide range of disciplines are needed to study combustion, which in turn has led to significant specialization within the field. Researchers developing chemical kinetic mechanisms, for example, will have expertise in quantum mechanics, computational chemistry, spectroscopy, and a number of zero- and one-dimensional experimental techniques, but they are unlikely to be experts in turbulent fluid mechanics or multiphase flow problems. Similarly, researchers with expertise in turbulence modeling or multiphase systems may have expertise in the development of sophisticated numerical algorithms but are unlikely to have any expertise in the development of chemical mechanisms or transport properties. Progress in each of these specialties would be more rapid and targeted if each of those communities had ready access to the best data, models, and software from the other specialties.

Historically, the linkages between the different subfields of combustion have been fairly weak. In particular, the linkages have been minimal between researchers investigating fundamental molecular-level properties such as kinetics and transport and researchers studying continuum behavior of flames. This historical structure of the field reflects the types of tools available to combustion scientists. Simulation capabilities for modeling reacting flows for most problems were limited to relatively simple (e.g., one- or two-step) chemical mechanisms that approximate (often poorly) the basic energetic and flame-propagation properties; computations of realistic flames with detailed kinetics were simply not feasible. Over the past decade, the advent of HPC and advances in numerical algorithms have changed this, and it is now feasible to simulate realistic turbulent flames with detailed kinetics for simple fuels (Chen, 2011; Lu and Law, 2009). Moreover, new quantitative diagnostic methods are being developed at all scales (Kohse-Hoinghaus et al., 2005), from individual reactive encounters, to controlled molecular ensembles, to in situ studies inside combustion chambers. As these measurement techniques are improved, much better characterization of detailed flame and ignition behavior will provide the data needed for the accurate validation of new simulation capabilities.

The changes in the tools available for studying combustion—combined with the need to address the challenges posed by developing cleaner, more efficient systems for new, alternative fuels—are driving a need to change the way that combustion research is being done.

COMBUSTOR CONFIGURATIONS

Practical combustion devices are usually complex systems in which the actual chemical reactions are only one part. To predict and evaluate combustor performance using a computer model, all of the processes that contribute to the overall system must be understood and included in the model, ideally with well-quantified fidelity. For safety concerns, fuel is almost always stored separately from the oxidizer, and they are not mixed at the molecular level until they are about to be burned. Therefore, the mixing of fuel and oxidizer either at the engine process level or within the flame structure is an essential process in the full combustion system, and in many cases the overall performance of the combustion system depends on the details of this mixing. Fuels may be initially in the gas, liquid, or

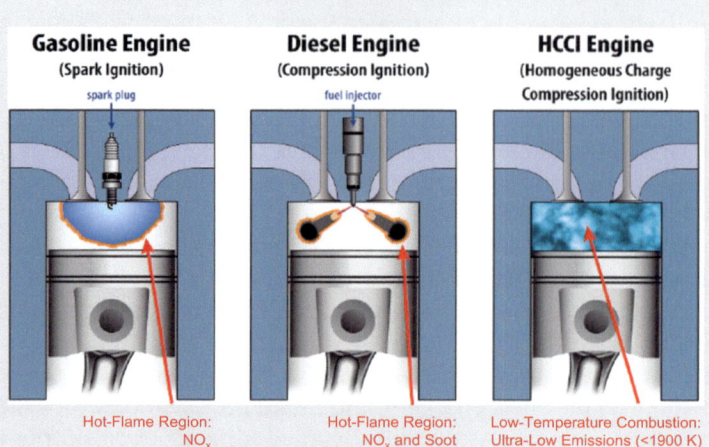

FIGURE 3.1.1 Internal combustion engine concepts.

BOX 3.1
Current and Future Internal Combustion Engine Concepts

In a typical gasoline engine, a premixed fuel-air mixture with just enough air to burn all of the fuel injected is compressed and spark-ignited (SI) at a specified, optimal time. The high temperature resulting from the stoichiometric combustion leads to significant NO_x formation. However, the NO_x, along with other pollutants formed, can largely be removed by the three-way catalyst aftertreatment system on all SI engines on the road today. The power output (load) of an SI engine is controlled by restricting (commonly called throttling) the amount of air drawn into the engine and injecting correspondingly less fuel. Overall, the throttling and low compression ratio of an SI engine result in a significantly less fuel-efficient engine than a diesel engine. In part, the lower efficiencies can be overcome with more recent developments that use direct injection of gasoline into the cylinder to eliminate the throttling losses. This option adds substantial flexibility toward ultralean combustion but at the expense of more challenging exhaust gas aftertreatment systems.

Direct injection can also be an enabler for homogeneous charge compression ignition (HCCI) engine technology, a new combustion strategy under investigation that approaches the high fuel efficiency of a diesel engine while producing very low NO_x and soot emissions—so low that there is the potential to meet the 2010 emissions standards without exhaust aftertreatment. As in an SI engine, fuel and air are premixed; however, combustion is started by a compression-ignition process similar to a diesel engine. Many challenges exist for this promising ultraclean combustion strategy before use in high-efficiency engines can be achieved. The challenges include robust methods of controlling the timing of ignition for optimal performance, expanding the usable load range, controlling the engine through transients, and determining the optimal fuel characteristics. Overcoming these challenges, especially in a diversified-fuel-source future, requires a vastly improved science base around the fundamentals of fuel ignition, combustion, and emission formulation chemistry, and fuel spray and turbulent fuel-air mixing processes, all at the high temperature and pressure conditions relevant to engines. These engine types are shown in Figure 3.1.1.

SOURCE: Adapted from www.er.doe.gov/bes.reports/files/ReproCTF_rpt.pdf.

solid phase, so fuel and air mixing can be a multiphase process. If the fuel is introduced into the combustion system in the liquid phase, such as in a diesel engine, then the combustion simulation must include a treatment of a liquid fuel spray and its vaporization and subsequent mixing with air. If the fuel is a solid particulate, a multiphase entrainment model is needed to characterize the system.

The combustible fuel and oxidizer mixture must then be ignited, and different practical devices accomplish this ignition in different ways. In some cases, such as diesel or homogeneous charge compression ignition (HCCI) engines, ignition is produced by the compressive heating of the reactants, whereas in other systems such as a gasoline engine, a spark or other ignition system is employed. These concepts are illustrated in Box 3.1. Once the fuel and oxidizer have been ignited, chemical reactions convert them into chemical products, and many different types of combustion can be encountered. Overall combustion can take place by means of nearly simultaneous fuel oxidation as in HCCI engines, but in most cases, combustion occurs through the propagation of flames. Some engines, such as spark-ignited gasoline engines, feature premixed flames, and other systems, including diesel engines and aircraft gas turbine engines, combine features of premixed and non-premixed flames—the latter being the central feature of most industrial burners. Reciprocating engines, found in most internal combustion engines, operate as repetitive time-dependent systems in a cyclically changing environment, whereas turbines used in aircraft and for power generation and industrial burners are statistically stationary turbulent flames (see Box 3.2) in which oscillation instabilities related to acoustic flame interactions are important phenomena. Each type of combustion system or flame requires its own solution algorithms, models, and data analysis.

In every important combustion system, the hot gas products transfer heat to do useful work, either by moving a piston or turbine blades or by heating a working fluid such as water in a boiler or furnace. In some large-scale systems, such as furnace combustion, building fires, and wildfires, radiation constitutes an essential long-range mode of heat transfer, and as such it must be included in the simulation. Furthermore, not all of the fuel in a combustion reaction is converted to chemical products such as CO_2 and H_2O, and incomplete fuel consumption is a primary source of carbon monoxide and other toxic emissions from combustion. In addition, the high temperatures produced by combustion convert molecular nitrogen (N_2) into oxides of nitrogen (NO_x), which are main components of photochemical smog. In some combustors, the incomplete or inefficient mixing of fuel with oxidizer can lead to the production of soot, which is produced by means of a complex sequence of chemical reactions that starts with gaseous species and steadily build larger molecules until they

BOX 3.2
Aircraft Emissions Produced by Combustion of Fuel and Air

Proposed high-efficiency engines will have combustors that must operate at extreme pressures (~50 atmospheres) and inlet temperatures (~950 K). Few facilities with necessary flow rates are available for testing such hardware and are very expensive to operate. Combustor modeling has assisted significantly in the design of recent-generation engines and can be expected to be even more critical at the extreme conditions in cycles of high-efficiency engines.

To enable such modeling, accurate simulations of phenomena such as fuel injection, spray and vaporization, fuel and air mixing, mixing and combustion in swirling and jets in-cross flows, heat transfer, fuel chemistry, and heat release rates in flows with Reynolds numbers on the order of 10^6 are required. Such capabilities are needed to ensure proper combustor performance by enabling quantitative predictions of flow splits and pressure drops, liner temperatures, exit temperature profiles at high power, emissions (gaseous and particulates) at various power settings, ignition, pre-ignition and altitude relight, lean blow-off, and acoustics. Furthermore, fuel effects need to be simulated—or experimentally validated—to ensure safe operation with new synthetic fuels. Testing new combustor designs through the use of advanced and validated modeling methods leads to significant savings. The operation of such an engine is shown in Figure 3.2.1.

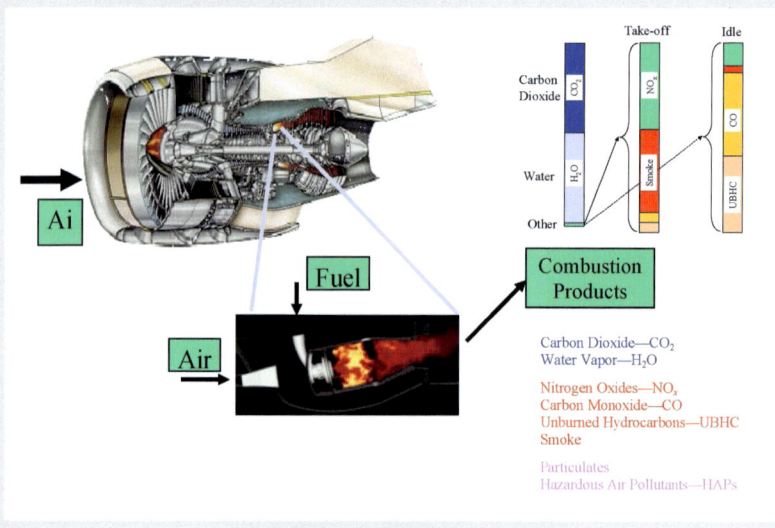

FIGURE 3.2.1 Turbine engine combustion.

become so large that they become solid-phase particles. In some cases, such as in industrial furnaces, soot at an intermediate stage in the combustion process is a good thing, enhancing radiative heat transfer from the burning gases to the surfaces of the combustor where heat transfer to a working fluid is the goal of the combustion process prior to its subsequent oxidation. In other cases, such as in diesel engines, soot is harmful and contributes to emissions of air pollutants.

MODELS, SUBMODELS, AND REDUCTIONS

One goal of combustion simulation is to provide system designers with a suite of simulation tools that can solve the relevant equations for all of the physical and chemical processes in an integrated model with quantified uncertainty. This full-system model thus has to include submodels for each of the processes that contribute to the performance of the combustor. Since the interactions between these subprocesses determine the overall performance of the combustor, each submodel must include the appropriate level of detail to be able to supply the overall model with enough information to simulate those interactions. The fidelity of the submodels needs to be categorized, and the impact of uncertainty in submodels on the overall simulation needs to be assessed. Some subprocesses do not contribute to every type of combustor; for example, in many gasoline-fueled, spark-ignition engines, the fuel is prevaporized prior to entering the combustion chamber, so a spray evolution submodel is not required. However, a spray submodel is essential for the simulation of combustion in a diesel engine, in which the liquid diesel fuel is injected directly into the combustion chamber and must vaporize before it can be burned.

Most practical combustion systems operate in a turbulent environment. Thus, a full-system model for a combustor is usually based on an underlying computational fluid dynamics (CFD) framework for turbulent flow that provides sufficient spatial resolution to describe the overall combustion process. In most cases, the physical combustor has a complex geometry, so this CFD framework must describe enough geometrical detail, usually in a fully three-dimensional setting. This geometric complexity can include the following:

- Fluid flows through moving intake and exhaust valves or through an injector,
- Swirling flows intended to enhance mixing between fuel and air,
- Flames propagating from spark plugs across a combustion chamber,
- Explosions occurring within some portions of a combustion chamber, and

- Radiation across a burner from a flame to tubes carrying secondary fluid to be heated.

The combination of three spatial dimensions and sufficient spatial resolution for each important subprocess often requires a very large computational grid, which leads to high computational costs. Each model for the subprocesses involved also can require extensive amounts of detail and therefore computer capacity. Since full-system detailed combustion models are prohibitively expensive for general use, the most common approach for full-system models is to combine simplified submodels for each of the important processes involved in a given combustion application. A balance is usually necessary, trading computational detail in each submodel against the need for an efficient computer solution, often combining many simulations at lower fidelity with a few simulations at higher fidelity to help bound uncertainties.

The starting point of a model for a combustion system is the description of the molecular properties of the fluids in the system. The physical and chemical parameters in a typical combustion model cover wide ranges, and the characteristics of these data are one of the primary reasons for the development of a CI for combustion. In particular, combustion systems cover temperature ranges from room temperature to levels of 3000 K in the products of fuel and oxygen flames, and from very low pressures of a few torr in laboratory flame studies to hundreds of atmospheres of pressure in diesel engines and some types of detonations. There can be many thousands of different chemical species when a hydrocarbon fuel is being burned, and each species has physical and chemical properties that are often complex functions of temperature and pressure. These include physical parameters such as equations of state, specific heat, heats of fusion and vaporization, ionization potentials and other spectroscopic quantities, frequency-dependent radiative opacities, molecular diffusivities, and others, as well as chemical parameters including elementary reaction rates with other species, with their activation energies and A-factors, heats of formation, and many other parameters.

For most applications, particularly those involving the modeling of a three-dimensional turbulent-flow, high-fidelity characterization of the fluid properties is not computationally feasible (see Box 3.3). For example, rather than using a chemical kinetic reaction model with 1,000 chemical species, a reduced chemistry model with perhaps 10 to 50 species needs to be used. The challenge is that it can be difficult for such a reduced chemistry model to reproduce the important features from a thousand-species chemistry model. Fortunately, extensive research in model reduction for chemical systems has shown that such reductions can be defined that are reasonably accurate and comprehensive within extensive parametric

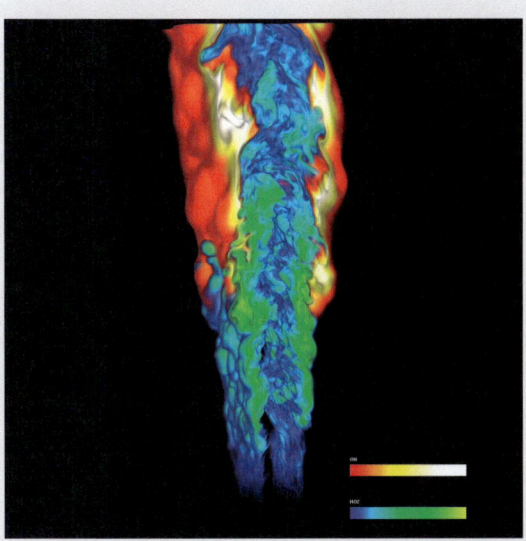

FIGURE 3.3.1 Direct simulation of a turbulent jet flame.

BOX 3.3
Stabilizing a Lifted Autoignitive Turbulent Jet Flame in a Heated Co-flow as Revealed by Direct Numerical Simulation

When a jet flame is lifted from the burner by increasing the fuel or surrounding air co-flow velocity, the flame can stabilize without a physical element to use for stabilization, and a lifted jet flame is generated. The flame can exist over a range of jet velocities until a critical velocity is reached and the flame blows out globally. Figure 3.3.1 shows a direct numerical simulation of a typical laboratory combustion configuration corresponding to a (small) turbulent, premixed, methane-air, slot Bunsen flame. The chemical model includes 12 reactive species and the computational domain is decomposed into 200 million cells.

Lifted flames are found in practical applications such as industrial burners for power generation, in which a lifted jet flame is utilized to reduce damaging thermal stresses to the nozzle material by minimizing contact between the flame and the nozzle. Lifted flames are also found in stratified combustion—for example, in direct injection gasoline engines, diesel engines, and gas turbines—where the fuel and oxidizer streams are partially premixed prior to combustion.

The position downstream of a fuel injector where a diesel fuel jet establishes a flame, the so-called stabilization point, influences the degree of premixing between the cold fuel and the heated air prior to combustion, which, in turn, affects the combustion and soot formation processes downstream. For example, soot levels decrease as the fuel and air streams are better mixed upstream of the stabilization point and can even be eliminated if sufficient premixing has occurred upstream. In a heated

(continued)

> **BOX 3.3 Continued**
>
> co-flow, the stabilization point is highly sensitive to the ignition quality of the fuel. For example, in a diesel jet flame the composition of the cool flame ignition intermediates and low-temperature heat release rate may help to stabilize the flame. However, cool flame chemical kinetics is relatively slow and may be modulated by turbulent strain and mixing. Therefore, fundamental knowledge of the mechanism by which a turbulent lifted flame stabilizes in a hot environment of vitiated gases will lead to the development of predictive models for the design and optimization of fuel-efficient, clean-burning combustion devices.
>
> ---
>
> SOURCE: See http://vis.cs.ucdavis.edu/~yuho/combustion/HO2+OH_8k.png. Accessed February 18, 2011.

ranges of operation (Lu and Law, 2009). Furthermore, reduced models can be patched together as operating conditions change, to provide even more realistic overall simulations.

There is, in fact, an entire discipline of mechanism reduction for chemical kinetics. Some of these approaches use an elementary approach by just removing chemical species and elementary chemical reactions that are relatively unimportant, whereas other, more complex approaches take advantage of the fact that strong coupling exists between the evolution equations for groups of chemical species (Prager et al., 2009; Liang et al., 2009; Hughes et al., 2009; Bykov and Maas, 2009). Other interesting approaches precompute representative chemical solutions and then save the answers, which can then be retrieved while carrying out a larger simulation, thereby saving the time required for the costly kinetics equations. The same concept of precomputing the expensive parts of a larger simulation has been used to simulate classes of turbulent flames, which can be visualized as a complex mixture of microscopic laminar flamelets. One can precompute an entire library of laminar flamelets and save the results, then retrieve the relevant flamelets from the library while simulating a turbulent flame as a superposition of these microscopic laminar flamelets (Pope, 1997). A combustion CI could provide optimal access to existing results and kinetic rate data, which would accelerate the optimization process and enable individuals in this community to provide better research tools to others. At the present time, this type of progress is extremely slow, proceeding through journal publications and conference meetings, with little or no central, common collection of relevant results and synthesis of results into widely useful common models.

From this discussion, it is apparent that database requirements vary from subcommunity to subcommunity. Databases required for kinetics modeling are tens of megabytes, whereas those for direct numerical simulation (DNS) calculations for complex geometries with modest kinetic submodels are already into the terabyte range—and they are only expected to grow.

In addition to the complexities of chemistry, fluid flow in most practical combustion systems is highly turbulent. In fact, turbulent flows are observed in many reactive noncombustion systems in nature, including the atmosphere, in the oceans, and in stars such as the sun. Turbulence modeling is thus a feature of many types of system simulations. In chemically reactive combustion systems, turbulence is essential because it increases the overall rates of fuel and oxidizer mixing and combustion, making it possible to burn a large amount of fuel in a short period of time to produce the power that is needed by various types of engines and burners. Turbulence is very difficult to predict and model, since it is a stochastic, highly variable phenomenon; descriptions of the reactive flow over an enormous range of spatial and temporal scales are required.

Historically, turbulence has been modeled by making a series of approximations (Peters, 2000; Poinsot and Veynante, 2005), averaging the characteristic turbulent fluctuations over time and space intervals that have been made progressively smaller as the capacities of computers have permitted over the years. An important current target of combustion modeling research is direct numerical simulation (see Box 3.3) of turbulent reacting-flow phenomena. For a given problem specification in terms of flow geometry and the choice of kinetics and transport models and other processes, DNS approaches fully numerically resolve all of the turbulent and flame scales in space and time. DNS calculations do not require models for turbulence; turbulent phenomena are handled directly by solving the Navier-Stokes equation so that deterministic physical and chemical predictions can replace probabilistic descriptions of spatial and temporal fluctuations, which require closure assumptions that are frequently intuitively posited. Needless to say, such direct numerical simulations require enormous computer resources to simulate even small and greatly simplified systems, and they present unique challenges for simulations as described in Box 3.3 and in Appendix C in this report. They nevertheless provide trustable insights that are available in no other way. In addition to chemical kinetics and turbulent flow, other areas amenable to study using a combustion CI include, but are not limited to, surface chemistry and catalysis and condensed-phase phenomena.

In conducting full-scale practical combustion simulations, however, it is not possible to resolve all of the scales needed to represent turbulence.

For practical systems, some type of model for turbulence is required. There are a variety of approximations, such as Reynolds-averaged Navier-Stokes (RANS) simulation and large-eddy simulation (LES) (see Box 3.4). RANS approaches, which are the traditional workhorse for engineering design simulations, attempt to capture the temporal average of a turbulent flow. LES approaches currently being developed are based on spatially filtering and incorporating a model for subgrid processes at finer scales than that of the filter. Typically, RANS is much less expensive than LES but offers less fidelity. Hence, there is a trade-off in the initial choice of turbulence modeling approach.

In nonreacting fluid flows, the choice of turbulence model depends on the fluid dynamical scales to resolve and the scales that require modeling. For reacting flows, the issues related to turbulence modeling are far more complex. Every subprocess that is being modeled in a combustion simulation requires a subgrid model for describing how that process interacts with turbulence. A model for mixing of fuel and oxidizer prior to combustion is needed. Models for turbulence-chemistry interaction are needed to represent flame behavior in a turbulent environment. When multiphase and radiation are important processes, submodels for their interaction with turbulence are also required.

Each of these submodels tries to represent very complex phenomena that are beyond the capabilities of existing computing. Thus, the improvements that are made with each advance in computing enable the inclusion of important phenomenology that was previously neglected. Such improvements are not simply incremental refinements but capture important details that require greater computing power. As more of the relevant phenomena can be directly captured in a simulation, fewer effects need to be incorporated in a subgrid model, resulting in higher-fidelity simulations. In addition, a fundamental reason for continually improving the submodels is so that the state-of-the-art versions can be used to evaluate coarser approximations. Highly refined models serve a critical role for running sensitivity analyses that contribute to the understanding of what phenomenology, and at what degree of precision, must be retained when a submodel is incorporated into a larger model to calibrate the fidelity of coarser models and estimate their associated uncertainties.

The researchers developing new fuels and combustion systems need to be able to simulate the behavior of proposed new devices and products under a wide range of operating conditions, in order to properly assess and improve the designs. Each of these simulations requires as inputs chemistry data relevant to the fuel of interest and some reacting-flow submodels (e.g., for turbulent mixing and radiative heat transfer). Thus, at a minimum the CI needs to carry information "up the scales," in the direction shown in Figure 3.1. Ideally the CI would enable feedback that

BOX 3.4
Large-Eddy Simulations

Computer simulations that are targeted for engineering applications require turbulence models. This is especially true for the complex flows that occur inside the cylinders of internal combustion (IC) engines. The next generation of turbulence modeling is called large-eddy simulations (LES). This name emphasizes the fact that the simulations can capture more of the important large-scale flow features than previous modeling approaches could. A generic name for these large-scale features is an "eddy," which qualitatively describes a swirling motion that is characteristic of turbulent flows. Thus, LES models offer enhanced representations of turbulent eddies and therefore better predictive capabilities. This is illustrated in Figure 3.4.1.

There are several requirements for LES: faster computers, improved numerical algorithms, and improved physical models. Faster computers are required so that more detail can be incorporated into the simulations. This requirement is progressively overcome with the rapid increase in computer capabilities. The requirement for more accurate, next-generation physical models involves more effort. The purpose of these models

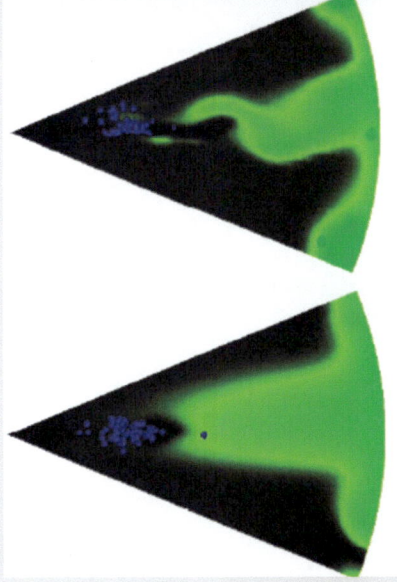

FIGURE 3.4.1 Snapshot showing combustion around one liquid fuel jet (injection from left to right) in a heavy-duty diesel engine. The top figure shows the experimental results, with the combustion displayed in green and the liquid fuel in blue. The middle figure shows an LES simulation of the same process, and the bottom figure shows a simulation using previous-generation turbulence modeling Reynolds-averaged Navier-Stokes (RANS) simulation. Note that LES captures the wavy structure of the combustion region, whereas RANS lacks resolution and shows only a smeared region for the combustion.

(continued)

BOX 3.4 Continued

is to represent the effects of small eddies—eddies that are too small to be captured in the simulation but still impact the large-scale dynamics.

A key feature of LES models is that they feature low dissipative errors and thereby achieve higher fidelity for the large-scale flow dynamics. Relatively simple LES models have been successfully developed for simple flows. However, for IC engines, with the added complexity of liquid fuel sprays and combustion, the models need to be more sophisticated. This is usually achieved with additional modeling variables and increased complexity.

SOURCE: Adapted from www.er.doe.gov/bes.reports/files/ReproCTF_rpt.pdf.

could carry information "down the scales" as well. This would allow simulations at the continuum level to be used for the verification of chemical mechanisms. Sensitivity analysis and uncertainty quantification at coarser scales can be used to identify research needs at finer scales. Integrating information flow across the scales would represent a fundamental change in combustion research. The proposed combustion CI can build on

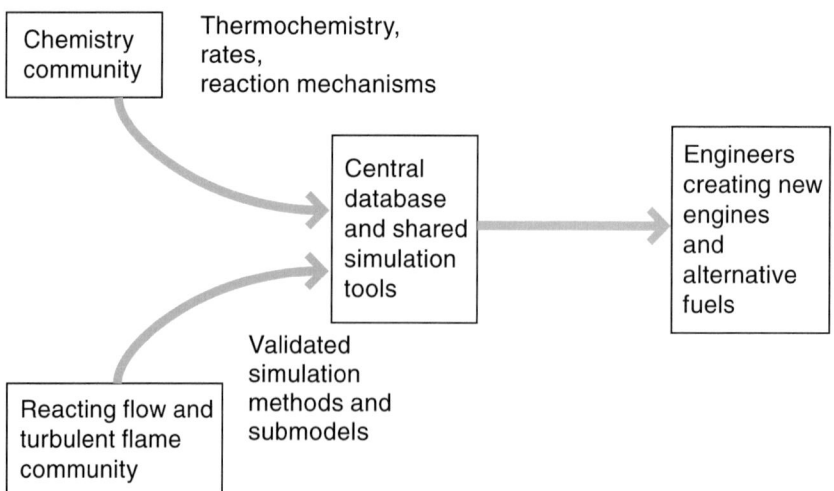

FIGURE 3.1 The cyberinfrastructure will collect the information and submodels needed by engineers to simulate novel combustion systems and alternative fuels under all operating conditions.

existing multiscale research. Two examples are multiscale simulations of nanoparticle formation from combustion sources (Viola and Voth, 2005) and flamelet modeling of turbulent combustion in which parameterized solutions from simple laminar flames are used to model the small-scale processes in turbulent flames (Peters, 1984).

DATA AND DATA FLOW

All of the data discussed above, spanning the range of conditions encountered in combustion systems, need to be made available to the combustion research community. However, it is frequently difficult to find all the data that one needs, and in some cases there can be quality-control problems. Although there have been data representation efforts—such as the representation of kinetics data at the Chemical Kinetics Database on the Web at kinetics.nist.gov, and the representation of molecular properties at the National Institute of Standards and Technology Chemistry WebBook at webbook.nist.gov—some data are very sparse or unavailable, because experimental measurements can be difficult or unrewarding. In addition, some data, both experimental and computational, are not sufficiently well characterized to meet the needs of the larger combustion community. However, the CI alone will not be sufficient. The community must pay attention to data reuse, including that of metadata, documentation, standards, and registries.

Species-Based Data

A common core of physical and chemical data is required for most combustion applications. These include specific heats, entropies, and enthalpies of formation; molecular transport coefficients; heats of fusion and vaporization; and other quantities that are specific for each individual chemical species involved in kinetic mechanisms, which are used to evaluate macroscopic variables such as the equations of state. Other species-specific quantities of importance include ionization potentials, radiative opacities and cross sections, and other spectroscopic parameters. Some data needs are specific to particular submodels, whereas others are needed for almost every submodel. For example, radiative opacities affect only the propagation of radiation within a combustor, but specific heats and pressures are needed for all cases. These data are determined in a variety of ways, and some can be provided by more than one physical model (Ritter and Bozzelli, 1991; Muller et al., 1995). Most thermochemical data are based on the "bond additivity" approach of Benson (Benson, 1976), which conceptually builds a molecule by combining all of the contributions from the atoms in the molecule and the bonds between them. However, these "groups" are evaluated in different ways, which result in similar but not identical values. There is a common data format for the results, based on a convenient approach from the Department of Defense's Joint Army-Navy-NASA-Air Force (JANNAF) tables (McBride et al., 1993). Almost every computational combustion researcher in the world uses thermochemical data tabulated in the JANNAF format, which is a good example of how standardization provides significant economies of time and effort. However, there is at present no ongoing effort to provide convergence of the various chemical models that generate JANNAF data into a single, consensus set of thermochemical data.

Chemical Databases

A central component of a combustion simulation is the submodel for combustion chemistry. For coupled chemical kinetic models, one differential equation must be solved for each chemical species in the reaction mechanism, and these differential equations are coupled through their mutual chemical reactions. While a hydrogen oxidation mechanism may be as small as 8 to10 species and is relatively easy to solve numerically, a mechanism for the important diesel reference fuel n-hexadecane (n-$C_{16}H_{34}$) can include 2,100 chemical species and 8,300 individual chemical reactions (Westbrook et al., 2009). Since chemistry submodels, especially those that compute overall rates of reaction and heat release, can include hundreds or thousands of different chemical concentrations, these databases can be extremely large.

Each chemical species adds a new differential equation to the overall combustion model; consequently, as the complexity of the chemical model grows, it can easily dominate the computational cost for a simulation. One approach to dealing with this chemical complexity is to reduce the dimensionality of the system. For example, for large kinetics models, many simulations are made on simplified, one-dimensional spatial grids that have several hundred spatial zones where large chemical mechanisms can be treated numerically. The best example of such one-dimensional models is the premixed laminar flame, which is widely used to study combustion kinetics for many fuels. Many other numerical studies, including a large variety of ignition problems as well as studies of oxidation in jet-stirred reactors and flow reactors, can be carried out under spatially homogeneous (i.e., single zone) conditions with no significant sacrifice of chemical generality.

Although these types of simulations can provide important insights into the behavior of the kinetic model, more realistic combustion simulations cannot treat a complex fuel with a comprehensive chemical mechanism. Thus, this chemical complexity motivates a need for reduced mechanisms, where the trade-off of using a simpler mechanism versus its reduced cost is well characterized. The CI must provide not only comprehensive mechanisms for a given fuel but also a hierarchy of reduced mechanisms for which the trade-off of fidelity versus computational complexity is well characterized.

The data needed to develop the high-fidelity chemical kinetics models represent a synthesis of data from a broad range of disparate sources. For example, the molecular enthalpies needed to simulate the combustion of an alternative fuel are extracted by combining information from spectroscopy, calorimetry, synchrotron photoionization, vapor pressure, reversible reaction kinetics, and mass spectrometry experiments with the results of quantum mechanical calculations. Each of those types of data is generated by different research communities, whose members attend different conferences and publish in different journals. When a new enthalpy value is reported, it is often discovered that there is a need to supplement previous measurements or calculations using new techniques in order to improve the precision or resolve technical issues with the older data. The CI needs to gather and disseminate all these disparate types of information, and therefore it must develop interfaces that are useful for each of the subcommunities in addition to the tools that facilitate the integration of these heterogeneous data (see Figure 3.2).

The combustion community is broken up into disparate subcommunities, each specializing in different types of data and methods. Combustion chemistry requires data from all of these communities. The CI will need to interface with many different subcommunities and handle very heterogeneous data. Because of the quantity and variety of data, data-curation services will be an important component of the combustion CI. These services maintain data integrity and enable data discovery, retrieval, and use over time through a range of activities for identification, selection, authentication, representation, preservation, and other processes that span the entire life cycle of data from production to archiving and reuse.

Continuum-Based Data

At its core, combustion is the synthesis of chemistry and fluid mechanics. The previous subsection discussed the role that the CI could play in enhancing progress at the molecular level. A CI could play a similar role at the continuum level, unifying the efforts of different research groups working on different aspects of the problem and providing a framework for dramatically improving the flow of information.

Broadly speaking, there are essentially three subcommunities whose research focuses on combustion at the continuum level. One of those sub-

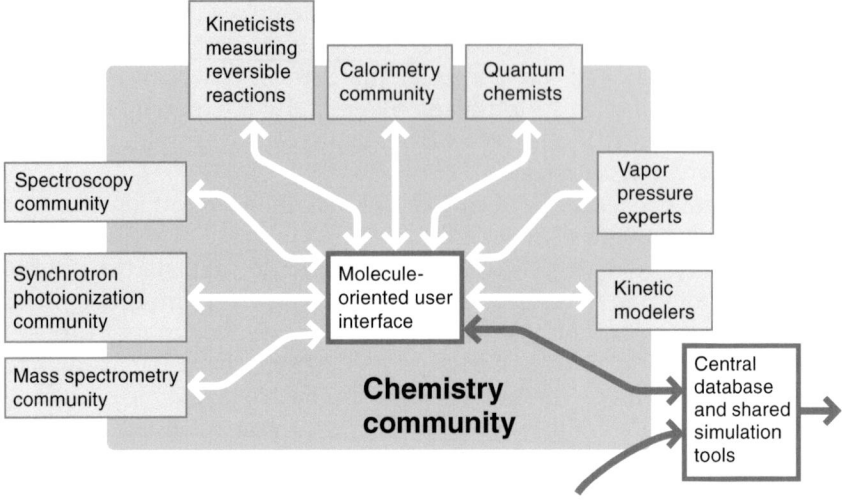

FIGURE 3.2 Expanded view of the chemistry community block from Figure 3.1.

communities consists of experimentalists whose research spans a range of problems, from simple laminar flames such as diffusion flames or vortex-flame interaction, to idealized turbulent flames in a laboratory setting, to in situ measurements in a realistic combustor.

The second of these subcommunities consists of researchers who develop and apply simulation methodology to study flame behavior at the continuum scale using a "first principles" approach; namely, solving the reacting Navier-Stokes equations. At the forefront of this subcommunity is a small number of research groups whose work focuses on harnessing the most powerful supercomputers to perform large-scale direct numerical simulations of turbulent flames. However, this subcommunity also includes groups using DNS approaches for less demanding turbulent-flame simulations as well as groups studying laminar-flame phenomena computationally.

The third subcommunity, whose efforts focus on continuum modeling, includes the groups that focus on the development of approaches to turbulence and combustion modeling and the application of those models to the simulation of realistic combustors. As with the other subcommunities, this group also spans a broad range of activities, ranging from those researchers developing high-fidelity LES approaches that capture significant features of a turbulent flow in detail, to researchers focused on improving the fidelity of simplified RANS-type approaches to full-system simulations, to engineers applying these tools as part of the design cycle to develop new combustion systems. A focal point of current LES efforts is the development of "multiregime" subgrid models for mixing and combustion that automatically treat the existence of mixed modes of combustion prevalent in advanced combustors operating at high pressure and low temperature, and in dilute conditions to maximize fuel efficiency and minimize emissions. For example, mixed combustion modes can include partially premixed flame propagation into low-temperature autoigniting mixtures for which traditional, purely premixed or purely non-premixed combustion models do not apply. Experimentation in these adverse environments is limited, and high-fidelity simulation is providing complementary data required to understand and model these complex combustion regimes. These regimes operate near combustion limits where intermittent phenomena—extinction, ignition, flammability limits—result in low-Damköhler-number flames and ignition phenomena characterized by strong sensitivities to variations in fuel chemistry and properties. Hence, there is demand for simulations that can differentiate fuel effects in fundamental "turbulence-chemistry" interactions encountered in practically relevant environments.

The CI requirements for the continuum community share a number of the issues with the infrastructure for molecular properties discussed above. The system would need to support a diverse set of data from a broad range of sources and to provide tools to facilitate the comparison of data from different sources. One example in this context would be tools for the direct comparison of experimental and simulation data. These types of comparisons would range from detailed comparison of laser measurement of nitrogen oxides (NO_x) in a diffusion flame, to a statistical comparison of local flame wrinkling in a turbulent premixed flame. Another type of comparison that would be required is the ability to perform comparisons between different types of models at different scales and fidelity. For example, the infrastructure should provide the ability to calibrate the fidelity of a coarse-grained turbulence model by comparing it with high-fidelity DNS data. For some of the data, quality issues are also important. For example, an experiment may not be adequately calibrated for it to be quantitatively compared to a simulation. Similarly, standards of validation and verification are needed to ensure the quality of simulation flame data. The committee believes that tools already developed by the combustion community, such as model versus data comparison tools developed by PrIMe,[1] will be made available to the combustion CI.

CI support for continuum data also presents some significant and

---

[1] See www.primekinetics.org. Accessed December 3, 2010.

unique challenges. The basic issue is that of the sheer volume of data associated with high-fidelity simulations. Time-dependent direct numerical simulations of turbulent flames being performed today are able to model billions of spatial zones for relatively simple gaseous fuels using either detailed or reduced chemical mechanisms. Typically, detailed mechanisms for hydrogen, syngas, or methane can be incorporated in DNS, whereas reduced mechanisms for larger hydrocarbon and oxygenated fuels—for example, n-heptane, di-methyl ether, butanol—accurately representing low-, intermediate-, and high-temperature kinetics at pressure (i.e., transporting approximately 30 to 80 reactive species) are required for feasibility in DNS on petascale supercomputers. These simulations currently generate data sets measured in the hundreds of terabytes and are only expected to increase in size over time. The data are characterized by velocity, temperature, and dozens of reactive scalar composition fields and additional hundreds of terabytes of particle data used to aggregate Lagrangian statistics in a turbulent flow. The generation of these data sets represents a significant computational investment on the world's largest supercomputers and, in many cases, is the only means of providing comprehensive temporally and spatially resolved validation data for model development. Hence, it would be extremely valuable to make these data sets available to the larger modeling community. Providing community access to this type of data raises a number of questions. Should there be an effort to collect such data at a central site, or should data be distributed, with individual data sets remaining at the facilities where they are generated? What types of tools are needed to catalog and access these data? If the data are not centralized, how does one deal with the security issues? What standardized file input/output formats and application programming interfaces are needed to facilitate extensible postprocessing and visualization software that operate on subsets of the data?

Similar to the GRIMech effort in kinetics, there has been one noteworthy example of an effort to organize a subset of the continuum combustion community, the Turbulent Nonpremixed Flame (TNF) Workshop. The TNF Workshop, organized in 1996 by Rob Barlow from the Sandia National Laboratories, is an open international collaboration of approximately 100 experimentalists and computational researchers focused on non-premixed and stratified turbulent flames. The TNF Workshop has established an Internet library of well-documented flames that are appropriate for model validation and for comparison of experimental data and modeling results. Although the TNF Workshop is limited to certain classes of flames and has not addressed the issues of providing community access to full-simulation data sets, it has served as a focal point for the community and is viewed as making a significant contribution to the study of non-premixed and, more recently, stratified turbulent flames. Thus far the scale of the experimental data is sufficiently small that data access, movement, and processing do not significantly hinder progress. The addition of high-fidelity simulation benchmark data and their comparison with experimental and RANS and LES models will require additional CI to accommodate the sheer volume of the simulated data.

Data Flow

All of the subcommunities involved in developing and verifying the molecular properties needed by the combustion community are in the chemical sciences, so they all communicate in a common scientific "language," and they all could be served by similar molecule-oriented user interfaces. Similarly, the different types of groups studying turbulent flames are from the fluid dynamics community and could be served by interfaces specific to their detailed research needs. However, the combustion CI must also facilitate information transfer between more disparate communities, and in some cases it would be appropriate to develop different interfaces for different communities. In particular, researchers studying continuum-level phenomena will require a substantially different interface to the chemical databases than that required by researchers studying kinetics of a particular molecule.

Another key ingredient is the need for the data flow in both directions. For example, researchers studying a turbulent flame may notice that there is a large discrepancy in the simulation predictions of a particular pollutant under certain reaction conditions, suggesting that something may be wrong with the chemistry model. The CI should allow for the rapid communication of this information back to the chemistry community (see Figure 3.3).

There is very strong data dependency in the combustion R&D system: a change in one number can require changes in many other numbers at all levels of the system. A dramatic illustration of this problem was the recent discovery that the published enthalpy of formation of OH was incorrect (Ruscic et al., 2001). This enthalpy number had been used to determine the enthalpies of many other species over the years, all of which need to change, and virtually every combustion simulation ever constructed is sensitive to this number, and so must be corrected (Ruscic et al., 2005). In the absence of a CI, all of these thousands of corrections must be performed by hand—a process expected to take many years. Clearly the new CI must correct this deficiency of the present information flow, making it practical to update the databases and the combustion simulators quickly whenever new information becomes available.

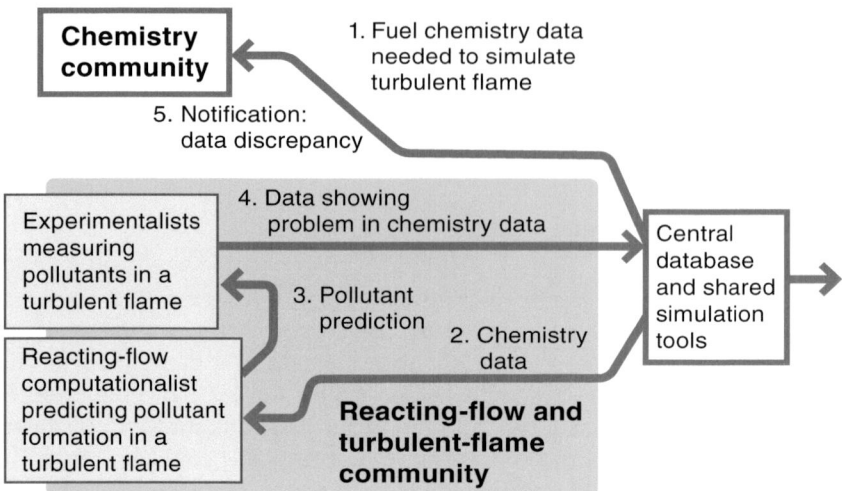

FIGURE 3.3 The cyberinfrastructure should facilitate forward and backward information flow between different communities, driving improvements in critical data and submodels.

The choice of enthalpy data in the examples above is just for illustrative purposes. There are many other types of data needed for combustion simulations, and most have similar issues (e.g., they are derived from heterogeneous sets of measurements and calculations; their numerical values are interdependent with other quantities; they necessitate information flow across disciplinary boundaries).

The development of submodels for many other processes could also be accelerated if coordinated partnerships, common data formats, and common software tools were available and used widely. Today, problems such as soot inception, growth, and oxidation are being pursued by many independent, individual groups, with minimal interconnection. Radiation transport in combustion devices, turbulence modeling, and many other data-rich submodels have the same characteristics as those outlined above. For example, the accessibility of detailed DNS data sets to the community developing turbulence models could greatly accelerate the development, validation, and verification of those models. The overall rate of progress toward solutions can be much faster, and the community can develop better capabilities with coordinated projects that are built around a centralized data-management and software tools core. This approach has never been attempted at this scale in the field of combustion, although in the few cases where this type of activity was pursued at a much reduced scope, it was very successful. Examples are PrIMe and CHEMKIN (see Appendix B in this report).

Cyberinfrastucture: A New Mode of Organization for a Community-Level Vision in Combustion Research

A CI-enabled combustion community will evolve in a two-tier framework organized into three quasi-homogeneous subcommunities (the fuel and chemistry research community, the reacting-flow and flame research community, and the industrial R&D community) under the umbrella of a superstructure for community-level coordination (see Figure 3.1).

To facilitate this transformation and to promote the fast, wide, and deep adoption of a CI by combustion researchers, it is envisioned that some of the current stakeholders of the research community will play an increasingly important leadership role. One such stakeholder is the Combustion Institute (www.combustioninstitute.org). The Combustion Institute is the premier international scientific society in combustion. With approximately 2,000 full-time members, it promotes the field of combustion science by organizing meetings, both national (the biennial U.S. National Combustion Meeting) and international (the biennial International Combustion Symposium), and by helping manage the premier journal publications in combustion science.

The Combustion Institute can play a pivotal role by providing the technical expertise, the integrity, and the legitimacy that an effort to build and sustain such a CI will require when interacting with funding agencies, individual researchers, and external communities. The Combustion Institute can work at the U.S. national level (through the joint U.S. sections of the Combustion Institute) or at the international level and thereby provide international coordination, should wide-scale CI efforts also be developed in Europe and/or Asia (in fact, the Combustion Institute can presumably promote the emergence of additional CI efforts in Europe and Asia). Finally, and perhaps more importantly, the Combustion Institute can further develop, articulate, and help promote the vision of a new combustion research community empowered by a community-wide CI. Other institutions that could undertake this role include the National Center for Supercomputer Applications at the University of Illinois at Urbana-Champaign and the National Institute of Standards and Technology.

REFERENCES

Benson, S.W. 1976. *Thermochemical Kinetics.* 2d ed. New York: Wiley.

Bykov, V., and U. Maas. 2009. "Problem Adapted Reduced Models Based on Reaction-Diffusion Manifolds (REDIMS)." *Proceedings of the Combustion Institute,* Vol. 32, pp. 561-568.

Chen, J.H. 2011. "Petascale Direct Numerical Simulation of Turbulent Combustion—Fundamental Insights Towards Predictive Models." *Proceedings of the Combustion Institute,* Vol. 33, pp. 99-133.

Frenklach, M. 2007. "Transforming Data into Knowledge—Process Informatics for Combustion Chemistry." *Proceedings of the Combustion Institute,* Vol. 31, pp. 125-140.

Hughes, K.J., M. Fairweather, J.F. Grifiths, R. Porter, and A.S. Tomlin. 2009. "The Application of QSSA via Reaction Lumping for the Reduction of Complex Hydrocarbon Oxidation Mechanisms." *Proceeding of the Combustion Institute,* Vol. 32, pp. 543-552.

Kohse-Hoinghaus, K., R.S. Barlow, M. Alden, and J. Wolfrum. 2005. "Combustion at the Focus: Laser Diagnostics and Control." *Proceeding of the Combustion Institute,* Vol. 30, pp. 89-123.

Liang, L., J.G. Stevens, and J.T. Farrell. 2009. "A Dynamic Adaptive Chemistry Scheme for Reactive-Flow Computations." *Proceedings of the Combustion Institute,* Vol. 32, pp. 527-534.

Lu, T.F., and C.K. Law. 2009. "Toward Accommodating Realistic Chemistry in Large Scale Computations." *Progress in Energy and Combustion Science,* Vol. 35, pp. 192-215.

McBride, B.M., S. Gordon, and M.A. Reno. 1993. "NASA Thermodynamics Database." NASA Technical Memorandum 4513. Cleveland, Ohio: Lewis Research Center, NASA.

Muller, C., V. Michel, G. Scacchi, and G.M. Come. 1995. "THERGAS: A Computer Program for the Evaluation of Thermochemical Data of Molecules and Free Radicals in the Gas Phase." *Journal of Chemical Physics,* Vol. 92, pp. 89, 1154-1178.

Peters, N. 1984. "Laminar Diffusion Flamelets in Non-Premixed Turbulent Combustion." *Progress in Energy Combustion Science,* pp. 319-339.

Peters, N. 2000. *Turbulent Combustion.* Cambridge, United Kingdom: Cambridge University Press.

Poinsot, T., and D. Veynante. 2005. *Theoretical and Numerical Combustion.* Philadelphia, Pa.: R.T. Edwards.

Pope, S.B. 1997. "Computationally Efficient Implementation of Combustion Chemistry Using in Situ Adaptive Tabulation." *Combustion Theory Modelling,* Vol. 1, pp. 41-63.

Prager, J., H.N. Najm, M. Valorani, and D.A. Goussis. 2009. "Skeletal Mechanism Generation with CSP and Validation for Premixed n-heptane Flames." *Proceedings of the Combustion Institute,* Vol. 32, pp. 509-518.

Puduppakkam, K.V., L. Liang, C.V. Naik, E. Meeks, and B. Bunting, 2009. *Modeling of a Gasoline HCCI Engine Using Model Fuels.* Society of Automotive Engineers Publication SAE-2009-01-0669. Warrendale, Pa.: Society of Automotive Engineers.

Ritter, E.R., and J.W. Bozzelli. 1991. "THERM: Thermodynamic Property Estimation for Gas-Phase Radicals and Molecules." *International Journal of Chemical Kinetics,* Vol. 23, pp. 767-778.

Ruscic, B., D. Feller, D.A. Dixon, K.A. Peterson, L.B. Harding, R.L. Asher, and A.F. Wagner. 2001. "Evidence for a Lower Enthalpy of Formation of Hydroxyl Radical and a Lower Gas-Phase Bond Dissociation Energy of Water." *Journal of Physical Chemistry A.* 105(1):1-4.

Ruscic, B., J.E. Boggs, A. Burcat, A.G. Scaszar, J. Demaison, R. Janoschek, J.M.L. Martin, M.L. Morton, M.J. Rossi, J.F. Stanton, P.G. Szalay, P.R. Westmoreland, F. Zabel, and T. Berces. 2005. "IUPAC Critical Evaluation of Thermochemical Properties of Selected Radicals, Part I." *Journal of Physical Chemistry Reference Data,* Vol. 34, pp. 573-656.

U.S. EIA (U.S. Energy Information Administration). 2007. *U.S. Crude Oil, Natural Gas, and Natural Gas Liquids Reserves, 2006 Annual Report.* DOE/EIA-0216. Washington, D.C. November.

U.S. Global Change Research Program. 2009. *Global Climate Change Impacts in the United States.* New York, N.Y.: Cambridge University Press.

Viola, A., and G.A. Voth. 2005. "A Multi-Scale Computational Approach for Nanoparticle Growth in Combustion Environments," in *High Performance Computing and Communications,* Vol. 3726, pp. 938-947, Book Series: Lecture Notes in Computer Science, L.T. Yang et al. (eds.). Berlin/Heidelberg: Springer.

Westbrook, C.K., Y. Mizobuchi, T. Poinsot, P.A. Smith, and J. Warnatz. 2005. "Computational Combustion." *Proceedings of the Combustion Institute,* Vol. 30, pp. 125-157.

Westbrook, C.K.,W.J. Pitz, O. Herbinet, H.J. Curran, and E.J. Silke. 2009. "A Comprehensive Detailed Chemical Kinetic Mechanism for Combustion of n-Alkane Hydrocarbons from n-Octane to n-Hexadecane." *Combustion and Flame,* Vol. 156(1), pp. 181-199.

4

Recommendations

The previous chapters in this report describe how modern information technology can be applied to speed the development of new combustion systems and alternative fuels, which would provide significant benefits to the nation's economy, to the environment, and to U.S. national security. Such technology would also accelerate the advancement of science and increase the effectiveness of many government-sponsored research efforts.

This chapter focuses on the particularly urgent need for development of a cyberinfrastructure (CI) for combustion, to ensure a more rapid and reliable data flow connecting three communities (see Figure 4.1):

1. *The fuel research community,* which is focused primarily on molecules and chemistry;
2. *The reacting-flow research community,* which is focused on understanding flame structure and combustion dynamics; and
3. *The industrial engine, combustion, and fuels research and development (R&D) community,* so that it can rapidly put new understanding of combustion to practical use for society's benefit.

Because of the different focus and organization of each of these communities, each poses different challenges for a CI.

For the fuels and chemistry communities, a CI must handle an extremely heterogeneous data set, including properties of individual molecules, reaction rates and transport properties, fuel compositions, and experimental data on the performance of different fuels under vari-

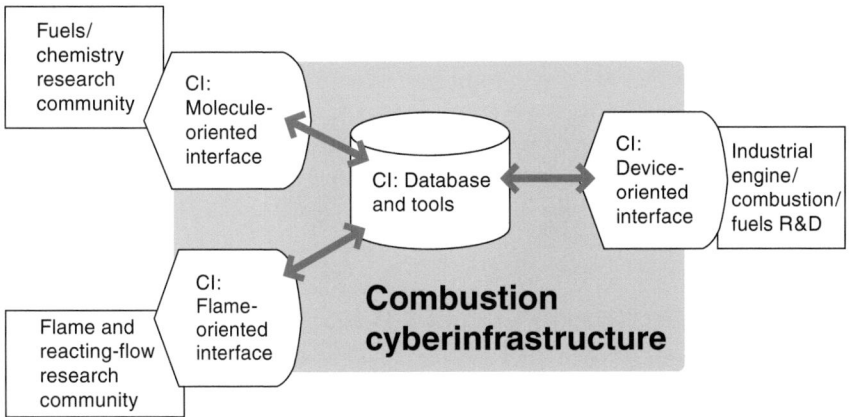

FIGURE 4.1 Overview of the needed technical components of a combustion cyberinfrastructure (CI), which will facilitate rapid data flow and improved coordination among various subcommunities contributing to research and development (R&D) in the areas of engines and fuels.

ous conditions. In this research area, which is dominated by chemists, information is usually organized by molecule. These data are being generated at a relatively slow rate by a very large number of researchers in a variety of disciplines scattered around the world. Much of the data has not yet been captured in usable electronic form, and extensive outreach to the combustion-chemistry community will be required to make the data more accessible and useful and to ensure that the data are up to date. Fortunately, in this area there is significant prior work and a history of cooperative community efforts to build on.

For the reacting flow community, the CI must handle very large volumes of data, but the data are more homogeneous, and the computational and experimental data of greatest interest are generated by researchers at supercomputer centers, national laboratories, and universities. Although there are still significant community-outreach and data-documentation issues impacting the flame and reacting-flow research community, they are not as severe as for the fuels and chemistry research community. Many of the flame and reacting-flow community's challenges here are technical: the mechanics of how to share and use extremely large data sets, how to ensure access to the validation data without compromising security, and so on. In this field, information is organized according to flame type and computational fluid dynamics (CFD) methodology.

To be most valuable to industrial engineers developing new engines, combustors, and fuels, the combustion CI must allow efficient access to a wide variety of data—from molecular properties to parameters of turbulence models, to data from a broad range of experimental and computational databases. The CI should also provide a variety of software tools for accomplishing common tasks, such as viewing and annotating the data, running simple simulations, and conducting comparisons with experimental data sets. In the industrial R&D community, information is organized by device type (e.g., jet engines, diesels, spark-ignited engines, and stationary turbines).

A CYBERINFRASTRUCTURE TO CONNECT COMBUSTION RESEARCH COMMUNITIES

Despite the significant differences among the three user communities—in the required interfaces and in the data and tools that each will find most interesting—it is essential that a combustion CI smoothly connect the three. This is because data on the particular fuel of interest are the key input to any reacting-flow simulation, whereas outputs from the reacting-flow simulations can identify which fuel components or properties have the greatest effects on performance in different types of engines and combustors. If the CI is working correctly, it is certain that the industrial R&D community will draw heavily on both the chemistry-oriented fuel community and on the CFD-oriented reacting-flow community.

As discussed in Chapter 3, the CI should facilitate two-way communication so that researchers will more quickly become aware of which data are currently available and which additional data are most needed. Because many prospective CI users are located at industrial R&D sites and at U.S. government facilities and because data on efficient engines and fuels are so economically and militarily valuable, the combustion CI would have to be designed to facilitate secure password protection on selected data and on access to computer resources, but without impeding efficient data flow. It is already the case that proprietary hardware and design drawings have been made available to universities for simulation and testing. It may be possible that design information that is not exact but close enough for validation of computational methods can be developed and used by the combustion CI.

Recommendation 1: A unified combustion cyberinfrastructure should be constructed that efficiently and effectively connects with and enables the movement of data and the sharing of software tools among the different research communities contributing to engine and combustion research and development.

The committee believes that the CI should be a platform that makes it easy for the community to contribute, integrate, and share software. Its resources should be largely developed by the CI's users and not entirely by its developers. Such an approach will also help create community buy-in and reduce costs.

In the sections below, the committee discusses and presents its recommendations with respect to each of the following areas:

- An organizational structure suitable for developing and maintaining the CI;
- The development of an implementation plan, including mechanisms for involving the broader combustion community;
- The educational role of the combustion CI; and
- Budgetary issues and sustainability.

The recommendations present detailed proposals for how this new CI should be created, introduced, and maintained. It is anticipated that building this CI will require an investment of tens of millions of dollars, and that the expected benefits are very much larger.

ORGANIZATIONAL STRUCTURE OF PROPOSED CYBERINFRASTRUCTURE

The proposed CI has three main requirements: (1) it must be unified and efficient, (2) it must interface effectively with the many subcommunities generating and using the data needed for combustion R&D, and (3) it must facilitate the validation of the submodels used in combustion simulations. These requirements and the experience of other CIs, discussed in Chapters 1 and 2, suggest a two-tiered organizational structure, with a central team and three individual outreach teams:

- *A central CI team* will be needed to devise the overall architecture and assemble and maintain the unified database and data-flow software that will connect all the communities.
- *The first outreach team will be focused on the molecule-oriented fuel research community*—developing data types, interfaces, and tools that will foster participation by each of the chemistry-oriented subcommunities and that will handle the highly heterogeneous chemistry data (from spectroscopy, calorimetry, kinetics, and so on).
- *The second outreach team will be focused on the reacting-flow and turbulent-flame community*—developing interfaces and tools for collecting combustion data and comparing them with full-model (or

submodel) predictions. This outreach team will also handle the specialized issues involving the very large data sets associated with petascale computer simulations.
- *The third outreach team will ensure that the combustion CI is effective in meeting the needs of the industrial engine and fuels R&D community.*

It is anticipated that many fuels researchers will interact with the first outreach team and that many mechanical and aerospace engineers involved in industrial combustion R&D will also interact with the second outreach team.

Recommendation 2: A centralized team will be needed to design and construct a unified, efficient combustion cyberinfrastructure in a timely fashion. At least three individual outreach teams should work closely with a central team: one outreach team connecting with the many chemistry-oriented subcommunities providing fuel data, one team connecting with the reacting-flow and turbulent-flame community, and one team ensuring that the cyberinfrastructure meets the needs of the industrial engine and fuels R&D community. These outreach teams will be responsible for interfaces, specialized software tools, and the development of formats and methods to handle different types of input data, and for the promotion of the new CI within their target communities.

IMPLEMENTATION PLAN

The development and deployment of a CI for combustion that effectively meets all of the requirements discussed above involve a complex, multifaceted, multimillion-dollar project. Detailed planning is necessary to ensure this important project is carried out in an efficient and timely manner. The implementation plan should include detailed specification for the core CI hardware and software, including detailed plans and funding levels with timelines for building the system.

This planning should be accomplished through a study that will also determine what historical data will be included in the CI databases, how those data will be put into the system, and the level of effort required to do so. There have been serious efforts in the past to collect and disseminate some of the data needed for combustion R&D, most notably by the National Institute of Standards and Technology, but also by the Jet Propulsion Laboratory, the American Petroleum Institute, the Joint-Army-Navy-NASA-Air Force Interagency Propulsion Committee, the Process Information Model, and the Turbulent Nonpremixed Flame Workshop. The planning study should indicate how the new CI will build on this prior work. Discussions with the curators of the data sets referred to above are needed from the outset.

Even more important than in the past, the planning study must determine how to interact most effectively with the various subcommunities involved in going forward. Early interaction with these subcommunities will help build a consensus on what is needed and will provide more detailed specifications on how the interfaces and software tools should work. Discussion groups and workshops may be an effective way to elicit input from the very beginning of the planning process. It is likely that some types of data could be most easily collected by cooperating with the journals publishing the original reports of the planning committee, but this might vary depending on the journal and the culture of the subcommunity involved. Ultimately, the success of the CI will depend on the level of engagement achieved with each community, so careful planning and effective execution of the CI's outreach component are essential. The implementation plan should also include contingency options for alternative budgetary levels.

Recommendation 3: Because of the many issues involved in the development and deployment of a CI for combustion, experts in several areas—chemistry data, reacting-flow simulations, engine and fuels R&D, software development, CI maintenance, data curation, deployment, outreach, and education—all need to be involved in the planning, design, and construction of the combustion CI.

A critical issue in the implementation plan is the question of how to get researchers to use the new system. It is likely that separate outreach teams will be needed for each of the subcommunities mentioned in this recommendation. In addition, the committee anticipates that once the CI is established, the added value that it provides to researchers will naturally lead to broad participation. Initially an incentive structure will facilitate early adopters. Federal grant agencies can provide incentives in many ways. In other fields, such as biomedical research, some agencies require data to be deposited in certain electronic archives prior to submission for journal publication. Such a requirement has been imposed by the National Science Foundation (NSF) for research grants beginning January 18, 2011:

> Beginning January 18, 2011, proposals submitted to NSF must include a supplementary document of no more than two pages labeled "Data Management Plan." This supplementary document should describe how the proposal will conform to NSF policy on the dissemination and sharing of research results.[1]

---
[1] See www.nsf.gov/bf/dias/policy/dmp.jsp. Accessed October 15, 2010.

Information regarding how a researcher interacts with the CI system, both as a user and as a provider of data and software, could be included in project reports and grant applications. This information could also play a role in the evaluation of progress in incorporating the use of the CI into the combustion community, along with journal publications and other metrics.

Recommendation 4: Federal research agencies responsible for funding combustion research should incorporate specific policies regarding the use of the combustion cyberinfrastructure into their progress reports and their grant processes. The incorporation of such policies will provide incentives to the combustion community and related communities for making the transition to the new system for handling and archiving valuable data.

A CYBERINFRASTRUCTURE AS AN EDUCATIONAL TOOL

Changes in Educational Programs

As demonstrated by the success of nanoHUB, a CI offers great opportunities to improve combustion education, including the following: better integration of combustion science and combustion engineering, providing stronger context and motivation for students; emphasis on the multiscale, multi-science nature of combustion, which positions it to be an excellent exemplar of state-of-the-art computational science and engineering; and better integration of combustion science with both computer and computational sciences, to expose students to a broader range of necessary tools and concepts.

It is noteworthy that instructors teaching combustion have a long history of working with open-source, cyber-based scientific data and software libraries, a history that started with the pioneering years of CHEMKIN (see Appendix B in this report). In combustion as in other engineering fields, the development of a cyber-based infrastructure drives profound transformations in the work environment and in professional practice; these transformations must be accompanied by corresponding changes in educational programs. Important ideas that should guide this transformation include the following:

- *A renewed emphasis on establishing stronger pedagogical ties between fundamentals (i.e., thermodynamics, chemistry, fluid mechanics, heat transfer) and applications (e.g., engine design).* By promoting unprecedented levels of integration, the CI can provide new ways to bridge the gap between different subcommunities (from commu-

nities of scientists who generate scientific ideas, to communities of engineers who generate design ideas and thereby express new scientific needs) and can enlist these subcommunities into a common framework.
- *The promotion of combustion as a multiscale discipline (from quantum and nanoscales to engineering device scales).* Combustion scientists and engineers will need to be exposed to an increasingly broad, cross-disciplinary, technical framework—for instance, to the concepts and tools of a multiscale approach to combustion. The framework of multiscale combustion includes molecular dynamics occurring at nanoscales, laminar-flame chemistry occurring at millimeter scales, turbulent-flame dynamics occurring at centimeter scales, and the overall engineering systems' performance characterized by scales on the order of tens of centimeters or more.
- *The integration of data science (a computer science topic) and scientific computing (a computational science topic) into the combustion curriculum.* The future needs of a cyber-based work environment will require the development of stronger ties among computer science, computational science, and domain science. For instance, combustion science and engineering students will need to be educated in the concepts and tools of the CI. These include a variety of information technology (IT) methods, such as software design, data structures, data visualization, and network architectures, in a distributed and heterogeneous (grid-like) environment, as well as a variety of computational science methods, such as numerical methods, parallel software design, and parallel computing optimization.

Educational Components

The committee envisions two main educational components in the combustion CI:

1. *The development of a combustion portal* similar to nanoHUB.org that would be a CI-enabled resource to the combustion learning community at large (including students, scientists, engineers, and policy makers). This resource would host data and software libraries and a portfolio of pedagogical Web-executable tools for focused classroom activities. In addition, it would host general instruction material for a reference combustion curriculum (including introductory tutorials, advanced courses, topical lectures, and so on). Finally, it would host introductory material for the general public, explaining the role and place of combustion science and engineering in meeting today's and tomorrow's energy challenges.

2. *The development of an advanced training program* that would be a CI-enabled resource to the combustion research community. This training program could take the form of specialized workshops, summer schools, Web conferences, and so on, and it would be aimed at disseminating the tools, standards, and methodologies of CI-enabled combustion science and engineering to the broader combustion research community.

The combustion portal will help develop and articulate a community-wide vision for combustion, including an enhanced integration of combustion science and combustion engineering. The advanced training program will help promote combustion as a multiscale, multi-science discipline and a leading application field for an enhanced CI with strong ties to computer and computational sciences.

These aspects of the combustion CI will facilitate the sharing of tools, data, and practices. However, new techniques for accelerating the process of journal publication, discussion, and critical dialogue, perhaps through mechanisms embedded in the CI, will be needed.

Recommendation 5: The combustion cyberinfrastructure should be designed to serve the chemistry and education communities as well as the research community, and to integrate these communities with advances in computer science research and education.

In its report *Revolutionizing Science and Engineering Through Cyber Infrastructure* (NSF, 2003, p. 17), the NSF elaborated on the contribution that a CI would make to education in any scientific or engineering discipline:

> These new environments can contribute to science and engineering education by providing interesting resources, exciting experiences, and expert mentoring to students, faculty and teachers anywhere. By making access to reports, raw data, and instruments much easier, a far wider audience can be served.

This committee believes that a combustion CI shares this promise.

BUDGETARY ISSUES

A precise estimate for the costs for implementing the proposed combustion CI cannot be determined until after the detailed implementation plan is developed. However, comparison with similar recent or ongoing CI efforts such as the CI at the National Center for Supercomputing Applications[2] (see Chapter 2 in this report) allows an estimate of the number of full-time employees that would be required. The centralized information technology functions of developing and maintaining the core database and data-flow software are estimated to require approximately 12 full-time-equivalent (FTE) employees, mostly computer programmers. The initial development of the specialized user interfaces, data types, and software; the related outreach to the many disparate subcommunities involved in combustion research; and the electronic capture of the historical data in each area will require a much larger workforce, which is a mix of domain experts (i.e., scientists and engineers who can communicate well with each subcommunity and who are familiar with the science issues) and computer programmers and data curators. This initial effort could easily require approximately 50 FTE employees for several years, although it might be decided to roll this effort out in stages, in part owing to the difficulty in rapidly identifying and hiring a sufficient number of qualified domain scientists and computer scientists. A team of this size, beginning a completely new project, would require a leadership and management team of perhaps 8 more FTE employees, yielding an estimate of about 70 FTE total employees at the project peak.

It is expected that a much smaller number of FTE employees would suffice to maintain the CI after it is established, but this ongoing maintenance is key to the success of the whole effort, and a revenue stream must be identified that will support the CI after it is initially constructed.

Many prior efforts at developing combustion CI have had little long-term value to the R&D community because they collapsed for lack of funding after the initial grants expired. The committee believes that a portion of the required ongoing revenue stream to maintain the combustion CI should come from industry. However, additional funds will certainly be required from the federal government, in part because so many of the likely users of the CI are in universities and national laboratories doing federally funded research, but also because the CI will no doubt be viewed as a "commons" rather than as a source of competitive advantage for an individual firm. The CI will be expensive, requiring new hardware, newly developed software, and a continuing support staff, all of which must be kept current. It is essential that the funding agencies plan for how to meet this long-term funding requirement from the very beginning. Otherwise, it is likely that society will not capture the value associated with this significant project. However, the CI should seek additional funding from sources other than the federal government, both for its own sustain-

---

[2]See ncsa.illinois.edu. Accessed September 20, 2010.

ability and to ensure that it is meeting the diverse needs of the combustion community. Such sources could include professional societies, industrial participants, and direct charges for the use of some of its facilities.

Recommendation 6: A fairly large short-term investment is required to achieve the benefits of a unified combustion cyberinfrastructure. Ongoing operations of this CI will require significant continuing funds. A failure to secure a continuing funding stream to maintain the CI will likely lead to the failure of the whole project.

The NSF, in its study *Investing in America's Future,* identified four goals to meet the national scientific needs: discovery, learning, research infrastructure, and stewardship. The combustion CI proposed in this report contributes to all of these goals; but, in particular, it addresses NSF's research infrastructure goal to "develop a comprehensive integrated cyberinfrastructure to drive discovery in all fields of science and engineering" (NSF, 2006, pp. 26-27). The development of a community-wide CI for combustion is a unique opportunity for the combustion community to reshape its structure and traditional modes of operation and thereby to achieve higher levels of integration and productivity. Such a community-wide CI is expected to drive a transformation of the combustion research community from a fragmented group of researchers and engineers characterized by light infrastructure and small research teams to an integrated community characterized by networked infrastructure and multidisciplinary research teams that function throughout the community. In the past, such a transformation toward a community-level integrated framework has been typically observed in research communities that share a large brick-and-mortar infrastructure (for instance, in the particle physics community, a community connected by the common need to use large-scale facilities such as particle accelerators). It is expected that a community CI will drive a similar transformation in the combustion research community; the drive and connections will, in that case, be provided by the common need to share data, software tools, computing resources, and personnel as well as the desire to bridge the gap between basic sciences and engineering applications.

REFERENCES

NSF (National Science Foundation). 2003. *Revolutionizing Science and Engineering Through Cyber Infrastructure: Report of the National Science Foundation Blue-Ribbon Advisory Panel on Cyberinfrastructure*. Arlington, Va.: National Science Foundation. Available at http://www.nsf.gov/cise/sci/reports/atkins.pdf. Accessed December 10, 2010.
NSF. 2006. *Investing in America's Future*. Arlington, Va.: NSF.

Appendixes

Appendix A

The GRIMech Model

Approximately 25 years ago, a project supported by the Gas Research Institute (GRI) in Chicago led to a common kinetic model for the combustion of natural gas, which is dominated by methane and has small levels of other gaseous components such as ethane, propane, CO, and H_2. A team of about a dozen combustion-chemistry experts collected extensive libraries of experimental and kinetic modeling papers and models of natural gas combustion. This team met several times a year, exchanged recommendations and opinions about which experiments were most reliable, carried out theoretical studies of important reactions and chemical species, and developed a fully detailed kinetic model—one that has been used for the past 20 years by researchers around the world. All of the information was assembled in a series of computer files, in a common format, with detailed descriptions of the data and methods used to optimize the resulting models, with significant curation of the entire system of data and evaluations.

The resulting model was called GRIMech (Frenklach, 2007), and over a period of years, several updates were released to the public. The benefits of the model included the consensus evaluations and recommendations of the panel of experts, the incorporation of modern data-curation and -evaluation processes in the model's development, and the broad range of applicability of the resulting model. Its availability for no cost was also an obvious factor in its wide acceptance and common use.

The degree of acceptance of this process led to the widespread use of GRIMech. It was particularly valuable for researchers who knew that their work required a reliable kinetic mechanism for methane or natural gas combustion chemistry, but who were not personally experienced or knowledgeable about combustion kinetics. GRIMech gave these people a tool that was ready to use, had been thoroughly tested by combustion-chemistry experts, and was freely available in a common and convenient format.

Although details of GRIMech are no longer considered to be leading-edge kinetic expertise, the model continues to be used extensively, for exactly the same reasons that made it successful 20 years ago—namely, easy and free access; thorough testing, evaluation, and validation; and common acceptance. This mechanism provided a combustion simulation tool of significant value that unified the international combustion community and accelerated combustion progress for 20 years. It is interesting to note that GRIMech shared most of these attractive features with the CHEMKIN software, described in Appendix B of this report. For more than 10 years, a large fraction of the combustion community used one or the other or both of these research tools in their daily efforts, and the community prospered and made significant advances in all types of combustion research, experimental and theoretical, as well as in computer modeling. These were two essential parts of an effective combustion cyberinfrastructure, and they form a template for a possible new cyberinfrastructure based on more modern software tools and chemistry models for practical petroleum-based fuels and biofuels for future combustion systems.

The GRIMech panel of experts envisioned extending this approach to larger and more complex hydrocarbons and other fuels, but the disappearance of the GRI as a research funding agency and the lack of other continuing support commitments made further extensions impossible. A combustion cyberinfrastructure would make these extensions possible. If the GRIMech history is any predictor, such a set of tools, enabled by a combustion cyberinfrastructure, would again have the same type of generally unifying results. It would also significantly accelerate the pace of progress in combustion research and lead to greatly improved combustion systems.

REFERENCE

Frenklach, M. 2007. "Transforming Data into Knowledge—Process Informatics for Combustion Chemistry." *Proceedings of the Combustion Institute,* Vol. 31, pp. 125-140.

Appendix B

CHEMKIN Chemical Kinetics Software

To provide insight into the role of various parameters and components of idealized combustion systems, software tools have been developed to simulate these systems. For example, models for premixed laminar flames started to appear in the 1960s at about the same time that research was revealing insights about chemical kinetic reaction mechanisms. Much of the new understanding about kinetics at the time was a result of the give-and-take between laminar premixed flame experiments and models.

The initial laminar flame models were inefficient and difficult to use, until a group of researchers (experimental and theoretical chemists and applied mathematicians) at the Sandia National Laboratories in Livermore, California, developed a significantly better flame model. That group also developed models for other common, idealized zero-dimensional combustion problems, including the plug-flow reactor, the jet-stirred reactor, the constant-volume reactor, and the rapid-compression machine. The group's full software package was entitled "CHEMKIN—A Chemical Kinetics Software Package."

The CHEMKIN software package provided a very convenient interface for specifying chemical kinetic reaction mechanisms, thermochemical data, and transport parameters, together with the computer software necessary to evaluate the various properties needed in the governing equations. Each of the individual system models was as good as, or significantly better than, any comparable models available at the time, and the authors made the entire package available to the general combustion community at no cost. In addition to the broad functionality of the software, distribution of the source code for CHEMKIN made it easy for others to modify the codes to suit any special needs of individual groups, greatly accelerating the growth in combustion-simulation capabilities in new directions.

This attractive combination of significant functionality, convenience, and zero cost rapidly made CHEMKIN the software tool of choice throughout the technical combustion world, and it rapidly became effectively an industry standard. This result had far-reaching and somewhat unexpected effects. It was no longer necessary for each research group to develop its own computational tools, especially the highly specialized models required for the challenging, stiff differential equations characteristic of chemical kinetic systems. It became easy for researchers anywhere to reproduce computational results carried out elsewhere. In addition, CHEMKIN enabled researchers with virtually no modeling expertise to build a computational component into their projects almost overnight. Most importantly, an individual researcher could go to another research organization anywhere in the world and quickly become completely functional because the new organization was using exactly the same modeling tools. Model improvements and new capabilities within the CHEMKIN family of models simultaneously provided the entire research community with the same new tools, with no cost or development expenses to anyone outside the core development group.

After 16 years, with minimal specified funding for its development, the CHEMKIN project at the Sandia National Laboratories ended, and the code itself was taken over by Reaction Design, a commercial company that has continued to improve the functionality and computational solvers in the CHEMKIN package (Reaction Design, 2009). However, Reaction Design started selling licenses for the software at costs that began to discourage some groups that were using these models. In addition, the codes were no longer open-source codes, so their flexibility in facilitating modifications was lost. As a result, although the CHEMKIN codes are still widely used, some users have changed to different modeling tools, while others continue to use the last free versions, which are now more than 10 years old.

The CHEMKIN history represents a case study of the enormous and far-reaching benefits made possible by a combustion cyberinfrastructure. There is enormous value in high-quality, practical, easily used software tools that simulate most of the common problems of importance to most research projects and that are constantly being refined and improved. In many cases, graduate students would learn how to use these computational tools during their master's and doctoral research and then carry that new expertise with them into their jobs, and current professionals could change jobs and continue to use the same software tools without further training. The same history also illustrates that the easy, free availability of the software is an essential piece of any cyberinfrastructure—the CHEMKIN codes are no longer an industry standard owing to increased cost, even though they have actually been improved following their privatization.

REFERENCE

Reaction Design. 2009. *Chemkin MFC-3.5*. San Diego, Calif.

Appendix C

Direct Numerical Simulations

Direct numerical simulation (DNS) is a computational approach specifically designed to capture key turbulence-chemistry interactions underlying the operation of modern combustors and, in particular, an approach that can discriminate the effects of differences in fuel properties. In DNS, all of the relevant turbulence and flame scales are resolved numerically using high-order, accurate numerical algorithms. Three-dimensional simulations with complex chemistry, and turbulence parameters within the realm of moderate Reynolds number flames, are of significant interest to the combustion-modeling community (academia, industry, and national laboratories) to provide data for the validation of key turbulence modeling concepts.

Many of the existing combustion models rely on conditioning variables, and it is feasible to assess the magnitude of the conditional fluctuations, leading to an understanding of the extent to which the thermochemical state can be parameterized by a reduced set of variables. Furthermore, it will be possible to assess the degree to which these variables can be predicted from the resolved or mean flow—for example, using presumed forms of the probability density function. For flamelet and transported probability density function models, Lagrangian tracking of fluid particles and flame elements in situ can be used to identify and understand the influence of unsteadiness, differential diffusion of species, and temperature on limit phenomena, including extinction and ignition phenomena. The particle-tracking statistics are also useful for understanding reactive scalar mixing rates and their departure from passive scalar mixing rates.

Once validated, the models can be used in engineering large-eddy simulation or Reynolds-averaged Navier-Stokes simulations to optimize the design of future devices.

As direct numerical simulations begin to move into the realm of laboratory-scale flames, DNS data can also provide valuable insights to the turbulent-flame experimental community. Experimentalists investigating turbulent-flame phenomena not only are limited by the scarcity of the data they are able to measure, but also often need to appeal to simplified theories to interpret measurements. For example, in turbulent premixed flame experiments, researchers typically assume that the flame locally propagates with the laminar-flame speed so that the turbulent-flame speed can be determined by measuring the wrinkling of the flame. This is an accurate approximation for natural gas flames but can lead to significant errors in regimes such as lean hydrogen flames. Simulation data from direct numerical simulations can be used to augment standard experimental diagnostics to obtain a more complete picture of turbulent-flame experiments.

A key issue in working with direct numerical simulations is the huge volume of data they generate—as much as 1/3 petabyte of data per realization on a 1 petaflop machine. Petascale computers are enabling DNS of turbulent reactive flows with approximately four decades of spatial scales and the incorporation of realistic chemical kinetics of small, single-component hydrocarbon fuels in three-dimensional turbulent simulations. The advent of exascale computing and beyond in the future will enable an even wider range of turbulence scales and the representation of more complex fuels at thermochemical conditions relevant to practical combustion systems, and the data volume and complexity will continue to increase.

However, such large and complex time-varying data sets—typically several hundred terabytes of raw data per simulation on today's 1 petaflop machines—pose serious challenges to gleaning physical insight from them and to sharing the data with the broader modeling community. Data-mining and database technology are needed to automate ways to identify patterns, to come up with reduced-order descriptions, to develop machine learning, and to visualize salient features in the data in order to reduce the sheer volume of data to be processed and transmitted.

Although only a few combustion researchers have the facilities, including access to high-performance computers, to carry out these enormous computations, access by the larger combustion community to the computed results, stored in a consistent and common format, is another important benefit that would be provided by a combustion cyberinfrastructure. These computed results represent a valuable resource that should be exploited by multiple researchers, but there is currently no clear way for such activities to be organized.

Appendix D

Chemical Kinetic Reaction Mechanisms

Chemical kinetic reaction mechanisms consist of two parts: (1) thermochemical data for chemical species that have been assembled into a systematic group to describe the combustion of a particular fuel, and (2) reaction-rate coefficients for the elementary chemical reactions in which those species participate. The most common format for such reaction-rate data is the modified Arrhenius form for the rate k:

$$k = AT^n \exp(-E_a/RT)$$

where the coefficients A, n, and E_a are tabulated for every reaction involved in the mechanism, T is the temperature of the burning system, and R is the universal gas constant. Recent mechanisms for practical diesel and biodiesel fuels can include nearly 3,000 chemical species and 11,000 elementary reactions (Westbrook et al., 2010; Naik et al., 2010). Different researchers develop detailed chemical kinetic reaction mechanisms of their own, using practices that can be called idiosyncratic but understandable. The most sensitive reaction rates for all of hydrocarbon combustion are those dealing with the smallest molecules, and while these rates have received an enormous amount of attention, there are still differences of opinion concerning the best ways to capture their rates into Arrhenius parameters.

The rates for the smallest molecules form the core of more complex kinetics mechanisms. Kinetic reaction mechanisms are best visualized as being "hierarchical" in structure (Westbrook and Dryer, 1984), based on a first level containing the kinetics for the smallest components, such as

H_2, O_2, CO, and CO_2, which are common to all hydrocarbon fuels. Submechanisms are then added for small hydrocarbons, such as methane (CH_4), formaldehyde (CH_2O), and methanol (CH_3OH), followed by fuels with two C atoms, and then by those with three C atoms. This is continued until the model size reaches that of the fuel being studied. Recent fuel mechanisms have been produced for fuels characteristic of gasoline, diesel fuels, and biofuels with as many as 16 to 18 carbon atoms (e.g., primary reference fuels, n-alkane, and methyl stearate, respectively). The numbers of chemical species and elementary reactions in a kinetic model increase rapidly with increased fuel molecule size:

- A model for hydrogen oxidation includes about 10 chemical species and 30 elementary reactions,
- A model for methane requires about 30 chemical species and 300 reactions, and
- A mechanism for n-cetane (n-$C_{16}H_{34}$) includes more than 1,200 species and 7,000 chemical reactions (Westbrook et al., 2009).

Starting with these core reactions and building on additional reactions for more complex fuels, each kinetic-model developer builds his or her own "house of cards" (Frenklach, 2007) that is similar to, but still not the same as, every other mechanism for the same fuels. For reaction rates with much less sensitivity than the above reactions, rates can vary much more widely from model to model. There is no real value in these differences from one model to another, but there is no motivation for convergence either. In the meantime, these multiple reaction mechanisms continue to exist, slowing the overall rate of progress in the direction of developing realistic reaction mechanisms for new, larger, and more practical fuel components. In addition, nonexperts who need usable and reliable kinetic mechanisms are confused and find it difficult to make choices from these multiple sources of models.

Experience with GRIMech (see Appendix A in this report) illustrates that it is possible for the combustion community to accept and thrive in a common, highly validated, and thorough reaction mechanism environment, but experience also has shown that when financial support for that kinetic model ended, the concept of a combustion community collective kinetic model disappeared.

It should be noted that, like many other features of a cyberinfrastructure, the chemical kinetic part would consist of a fairly complicated task in data management, visualization, and computer science in general, as well as in combustion kinetics. Large archives of experimental results would be needed as sources of model-validation data, and it would be necessary to provide such data in some sort of common data structure to enable models to access the data many thousands of times in validation studies. Common evaluations of relative confidence in those experimental data would have to be established, and software would be needed to carry out the various types of validation and other studies leading to consensus recommendations.

Teams of kinetics experts would be needed to develop analysis tools to arrive at common conclusions and recommendations, and optimization and evaluation tools would be required. Some of these were developed in the production of the GRIMech mechanisms, but the effort required to extend the approach to hydrocarbons in general would be an order-of-magnitude increase in necessary resources.

REFERENCES

Frenklach, M. 2007. "Transforming Data into Knowledge—Process Informatics for Combustion Chemistry." *Proceedings of the Combustion Institute,* Vol. 31, pp. 125-140.

Naik, C.V., C.K. Westbrook, O. Herbinet, W.J. Pitz, and M. Mehl. 2010. "Detailed Chemical Kinetic Reaction Mechanism for Biodiesel Components Methyl Stearate and Methyl Oleate." *Proceedings of the Combustion Institute,* Vol. 33, doi: 10.1016/j.proci.2010.05.007.

Westbrook, C.K., and F.L. Dryer. 1984. "Chemical Kinetics Modeling of Hydrocarbon Combustion." *Progress in Energy Combustion Science,* Vol. 10, pp. 1-57.

Westbrook, C.K.,W.J. Pitz, O. Herbinet, H.J. Curran, and E.J. Silke. 2009. "A Comprehensive Detailed Chemical Kinetic Mechanism for Combustion of n-Alkane Hydrocarbons from n-Octane to n-Hexadecane." *Combustion and Flame* 156(1):181-199.

Westbrook, C.K., W.J. Pitz, M. Mehl, and H.J. Curran. 2010. "Detailed Chemical Kinetic Reaction Mechanisms for Primary Reference Fuels for Diesel Cetane Number and Spark-Ignition Octane Number." *Proceedings of the Combustion Institute,* Vol. 33, doi:10.1016/j.proci.20.05.087.

Appendix E

Committee Meeting Agendas

MARCH 9-10, 2009
KECK CENTER OF THE NATIONAL ACADEMIES
WASHINGTON, D.C.

March 9, 2009

Closed Meeting
8:00 a.m.-10:00 a.m.

Open Meeting
10:15 a.m.-11:15 a.m. *Charge from Sponsor*
　　　　　　　　　　　Phillip Westmoreland, Program Director, Combustion, Fire, and Plasma Systems, National Science Foundation (NSF)

11:15 a.m.-12:15 p.m. *Office of CyberInfrastructure*
　　　　　　　　　　　Edward Seidel, Director, Office of CyberInfrastructure, NSF

12:15 p.m.-1:00 p.m. Working Lunch

1:00 p.m.-2:45 p.m. *Cyberinfrastructure for Metagenomics (CAMERA)*
　　　　　　　　　　　Jeffrey Grethe, University of California, San Diego (UCSD)

2:45 p.m.-3:00 p.m.	Break
3:00 p.m.-5:00 p.m.	*PrIMe* Michael Frenklach, University of California, Berkeley
5:00 p.m.-5:45 p.m.	Reception
6:00 p.m.	Working Dinner

March 10, 2009

8:00 a.m.-8:30 a.m.	Working Breakfast
8:30 a.m.-10:00 a.m.	*Panel Discussion with Leaders of Previous Workshops* Phillip Westmoreland, Program Director, Combustion, Fire, and Plasma System, NSF Douglas Talley, Air Force Research Laboratory, Edwards Air Force Base Arnaud Trouvé, Committee Member, University of Maryland
10:00 a.m.-10:15 a.m.	Break
10:15 a.m.-12:00 noon	Open Discussion with Speakers
12:00 noon-1:00 p.m.	Working Lunch

Closed Meeting
1:00 p.m.-3:00 p.m.	Closed Planning Discussion

JUNE 1-2, 2009
KECK CENTER OF THE NATIONAL ACADEMIES
WASHINGTON, D.C.

June 1, 2009

Closed Meeting
8:00 a.m.-9:00 a.m.

Open Meeting

9:00 a.m.-9:30 a.m.	*The Combustion Community* Mitchell Smooke, Committee Chair, Yale University
9:30 a.m.-10:30 a.m.	*The Cyberinfrastructure Community* Miron Livny, University of Wisconsin-Madison, and Dennis Gannon, Microsoft Research, Committee Members
10:30 a.m.-10:45 a.m.	Break
10:45 a.m.-11:45 a.m.	*Large Hadron Collider* Frank Wüerthwein, UCSD
11:45 a.m.-12:45 p.m.	Working Lunch
12:45 p.m.-1:45 p.m.	*ASC Alliance Centers* Thuc Hoang, Department of Energy (DOE)
1:45 p.m.-2:45 p.m.	*SciDAC* Walter Polansky, DOE
2:45 p.m.-3:00 p.m.	Break
3:00 p.m.-4:00 p.m.	*The NanoHUB Experience* Mark Lundstrom, Committee Member, Purdue University

Closed Meeting

4:00 p.m.-6:00 p.m.	Discussion

June 2, 2009

Open Meeting

8:30 a.m.-9:00 a.m.	Working Breakfast
9:00 a.m.-10:00 a.m.	*Sociological Aspects of Virtual Communities* Carol Palmer, Committee Member,

	University of Illinois, Urbana-Champaign
10:00 a.m.-10:15 a.m.	Break

Closed Meeting

10:15 a.m.-3:00 p.m.	Discussion

SEPTEMBER 30-OCTOBER 1, 2009
BECKMAN CENTER OF THE NATIONAL ACADEMIES
IRVINE, CALIFORNIA

JANUARY 19-20, 2010
KECK CENTER OF THE NATIONAL ACADEMIES
WASHINGTON, D.C.

The third and fourth meetings of the committee were held September 30 through October 1, 2009, at the Beckman Center, Irvine, California, and January 19-20, 2010, at the Keck Center in Washington, D.C.

Appendix F

Biographies of the Committee Members

Mitchell D. Smooke is the Strathcona Professor of Engineering at Yale University and the chairman of Mechanical Engineering and Materials Science. In addition, he holds a joint appointment in the Department of Applied Physics. Administratively, he has served as the dean of engineering, the chair and director of undergraduate studies of mechanical engineering, and the director of graduate studies for engineering. He received a B.S. in physics from Rennselaer Polytechnic Institute in 1973, a Ph.D. in applied mathematics from Harvard University in 1978, and an M.B.A. in management and finance from the University of California, Berkeley, in 1983. Before joining the faculty at Yale, he was a staff scientist at the Sandia National Laboratories in Livermore, California. Since 1996, Dr. Smooke has served as an editor in chief of *Combustion Theory and Modeling*. He is currently on the board of directors of the Combustion Institute and was a program co-chair for the 32nd International Combustion Symposium in Montreal, Canada, in 2008. Dr. Smooke has been the recipient of the Combustion Institute's Silver Medal and the Institute for the Dynamics of Explosions and Reacting Systems' Oppenheim Award. More recently, he received the Yale University Graduate School Mentor Award, the Yale College Teaching Prize for Science and Engineering, the Yale Engineering Sheffield Teaching Prize, and the Yale Science and Engineering Association Award for the Advancement of Basic and Applied Science. He is a fellow of the American Institute of Aeronautics and Astronautics, of the Institute of Physics, and of the Society for Industrial and Applied Mathematics (SIAM). Dr. Smooke has served on various technical boards and co-organized the U.S.-based SIAM meetings in computational combustion since 1991. His current primary research interests lie in the areas of computational combustion, energetic materials, chemical vapor deposition, and the numerical solution of ordinary and partial differential equations.

John B. Bell is senior staff mathematician and group leader for the Center for Computational Sciences and Engineering at the Lawrence Berkeley National Laboratory (LBNL). Prior to joining LBNL, he held positions at the Lawrence Livermore National Laboratory, Exxon Production Research, and the Naval Surface Weapons Center. Dr. Bell's research focuses on the development and analysis of numerical methods for partial differential equations arising in science and engineering. He has made contributions in the areas of finite difference methods, numerical methods for low-Mach-number flows, adaptive mesh refinement, interface tracking, and parallel computing. He has also worked on the application of these numerical methods to problems from a broad range of fields, including combustion, shock physics, seismology, flow in porous media, and astrophysics. He was the recipient of the SIAM/Association for Computing Machinery (ACM) Prize in Computational Science and Engineering in 2003, and he received the Sidney Fernback Award in 2005.

Jacqueline H. Chen is a distinguished member of the technical staff at the Combustion Research Facility at the Sandia National Laboratories. She has contributed broadly to research in petascale direct numerical simulations (DNSs) of turbulent combustion focusing on fundamental turbulence-chemistry interactions. These benchmark simulations provide fundamental insight into combustion processes and are used by the combustion-modeling community to develop and validate turbulent combustion models for engineering computational fluid dynamics (CFD) simulations. In collaboration with computer scientists, Dr. Chen and her team have developed algorithms and software for automated combustion work flow, in situ data mining and visualization of petascale simulated combustion data, and reacting-flow DNS software for hybrid architectures. She received the Department of Energy (DOE) Innovative and Novel Computational Impact on Theory and Experiment (INCITE) Award in 2005, 2007, and 2008-2010, and the Asian American Engineer of the Year Award in 2009. She is a member of the DOE Advanced Scientific Computing Research Advisory Committee and Subcommittee on Exascale Computing. She was the co-editor of the *Proceedings of the Combustion Institute*, Volumes 29 and 30, and a member of the executive committee of the board of directors of the Combustion Institute.

Meredith B. Colket III is a fellow in the Thermal and Fluid Sciences Department of United Technologies Research Center, where he has been employed for 31 years. He is an internationally known expert on the gas-phase kinetics of hydrocarbon fuels, on fuel preprocessing, probe phenomena, diagnostics for combustion-derived pollutants, and soot formation. Recently, he has led the development of new physics-based tools for the simulation of combustion phenomena, including emissions, in combustors and ignition of military augmentors. In addition, he has been helping develop technologies for the detection of toxic gases, fire suppression, and burner control. Dr. Colket has about 40 peer-reviewed publications, has been awarded 6 patents, and has been principal investigator/ program manager for a variety of government contracts, including four funded by the Air Force Office of Scientific Research (AFOSR) and two by the Air Force Research Laboratory. He also has been the recipient of two special awards and two outstanding achievement awards from the United Technologies Research Center; co-editor of the *Proceedings of the Combustion Institute*, Volumes 29 and 30; and has participated as member in a variety of panels for DOE, the National Science Foundation (NSF), the National Research Council (NRC), AIAA, and NASA. He was a colloquium chair for the 32nd International Symposium on Combustion. Dr. Colket is currently a member of the executive board of the Combustion Institute, an associate editor for *Combustion Science and Technology*, the task leader for the Jet Fuels Surrogate Working Group sponsored by AFOSR, and a member of a review committee for the Implications of Natural Gas Interchangeability for the California Energy Commission.

Thomas H. Dunning is the director of the National Center for Supercomputing Applications and holds the Distinguished Chair for Research Excellence in Chemistry. He was at the Los Alamos National Laboratory in 1973, first in the Laser Theory Group and then in the Physical Chemistry Group. Dr. Dunning was appointed group leader of the Theoretical and Computational Chemistry Group at the Argonne National Laboratory in 1978. Beginning in 1989, Dr. Dunning held many positions at the Pacific Northwest National Laboratory, becoming director of the Environmental Molecular Sciences Laboratory in 1994 and the first Battelle Fellow in 1997. He spent 2 years (1999-2001) in the Office of Science of the U.S. Department of Energy as assistant director for scientific simulation; there he was responsible for developing a new scientific computing program. He then went to the University of North Carolina at Chapel Hill as a professor of chemistry and was responsible for supercomputing and networking for the University of North Carolina System. In 2002, he was appointed director of the Joint Institute for Computational Sciences, Distinguished Professor of Chemistry and Chemical Engineering at the

University of Tennessee, and Distinguished Scientist in Computing and Computational Sciences at the Oak Ridge National Laboratory. Dr. Dunning joined the University of Illinois faculty in January 2005.

Dennis Gannon is the director of engagements for the eXtreme Computing Group in Microsoft Research. Prior to coming to Microsoft, he was a professor of computer science at Indiana University and the science director for the Indiana Pervasive Technology Labs. Dr. Gannon's research interests include cloud computing, large-scale cyberinfrastructure, programming systems and tools, distributed computing, computer networks, parallel programming, computational science, problem-solving environments, and performance analysis of supercomputing and distributed systems. He led several software projects for the Defense Advanced Research Projects Agency and the Department of Energy related to programming massively parallel systems. He has worked extensively with National Science Foundation (NSF)-sponsored interdisciplinary scientific teams on applications ranging from computational cosmology to predicting tornadoes and hurricanes. Dr. Gannon was on the executive steering committee of the NSF TeraGrid, where he helped launch the Science Gateways project designed to enable access to supercomputing to a larger audience of researchers. He also managed the TeraGrid Science Advisory Board. Dr. Gannon was the program chair for the Institute of Electrical and Electronics Engineers (IEEE) 2002 High Performance Distributed Computing Conference, the 1997 ACM International Conference on Supercomputing, the 1995 IEEE Frontiers of Massively Parallel Processing, and the International Grid Conference in Barcelona, 2006. He also served as general chair of the 1998 International Symposium on Scientific Object Oriented Programming Environments, the 2000 ACM Java Grande Conference, the ACM Principles and Practices of Parallel Programming Conference, and the ACM International Conference on Supercomputing. While he was chair of the Computer Science Department at Indiana University, he led the team that designed the university's new School of Informatics. For that effort he was given the school's Hermes Award in 2006. He has published more than 100 refereed articles, and he has co-edited 3 books. Dr. Gannon received his Ph.D. in computer science from the University of Illinois at Urbana-Champaign in 1980 after receiving a Ph.D. in mathematics from the University of California, Davis.

William H. Green is a professor of chemical engineering at the Massachusetts Institute of Technology. His research group focuses on the central problem of reactive chemical engineering: quantitatively predicting the time evolution of chemical mixtures. Dr. Green has received the C.M. Mohr Award for Outstanding Undergraduate Teaching in 2006, American

Chemical Society (ACS) Division of Environmental Chemistry Certificate of Merit in 2005, the ACS Division of Fuel Chemistry Richard A. Glenn Award in 2004, the Thiele Lectureship Award (Notre Dame University) in 2004, and the NSF CAREER award for 1999-2003. He was an NSF postdoctoral fellow in 1989-1990, a Darwin Research Fellow at Cambridge University during 1989-1990, a NATO postdoctoral fellow in 1989, an Amoco Foundation Fellow in 1987, and an NSF graduate fellow during 1983-1985.

Chung K. Law is the Robert H. Goddard Professor in the Department of Mechanical and Aerospace Engineering at Princeton University. His research interests are in combustion, propulsion, heat and mass transfer, energy, alternative fuels, and the environment. His research accomplishments have included his receiving the Curtis W. McGraw Research Award of the American Society for Engineering Education (ASEE) in 1984; a Silver Medal and the Egerton Gold Medal of the Combustion Institute in 1990 and 2006, respectively; the Propellants and Combustion Award, the Energy Systems Award, and the Pendray Aerospace Literature Award of the American Institute of Aeronautics and Astronautics in 1994, 1999, and 2004, respectively; the Heat Transfer Memorial Award of the American Society of Mechanical Engineers (ASME) in 1997; and outstanding alumnus awards from the University of California at San Diego in 2000 and the Hong Kong Polytechnic University in 2007. He is a fellow of the AIAA, ASME, and the American Physical Society (APS); a member of the National Academy of Engineering (NAE); a fellow of the American Academy of Arts and Sciences; the director of the DOE Combustion Energy Frontier Research Center; the director of the Center for Combustion Energy at Tsinghua University, China; and a former president of the Combustion Institute.

Miron Livny is a senior researcher and professor specializing in distributed computing at the University of Wisconsin-Madison. He has been a professor of computer science at Wisconsin since 1983, where he leads the Condor high-throughput computing system project. He is also a principal investigator and currently the facility coordinator for the Open Science Grid project. In 2006, along with Raghu Ramakrishnan, Professor Livny won the ACM's Special Interest Group on the Management of Data (SIGMOD) Test of Time award for his seminal work on distributed databases.

Mark Lundstrom is the Don and Carol Scifres Distinguished Professor of Electrical and Computer Engineering at Purdue University. His research and teaching focus on the theory, modeling, and simulation of nanoscale electronic devices. He co-founded the Purdue University Network Computing Hub project, an early example of cyberinfrastructure that delivered electronic device simulation services through the World Wide Web. He was the founding director of the NSF-funded Network for Computational Nanotechnology, which has developed a second-generation cyberinfrastructure that supports the U.S. National Nanotechnology Initiative. He now directs the Network for Photovoltaic Technology, a Semiconductor Research Corporation consortium of companies in the United States, Asia, and Europe created to fund university research in photovoltaic technology. He is a fellow of the IEEE, the APS, and the American Association for the Advancement of Science, and he is a member of the NAE.

C. Bradley Moore is professor of chemistry emeritus in the Department of Chemistry, University of California, Berkeley. His physical chemistry research includes molecular energy transfer, chemical reaction dynamics, photochemistry, and spectroscopy. Applications of his work are found in combustion and atmospheric chemistry, in chemical and molecular lasers, and in isotope separation. He is a member of the National Academy of Sciences and the American Academy of Arts and Sciences. He has served as chair and dean at Berkeley and as vice president for research at the Ohio State University and at Northwestern University.

Carole L. Palmer is a professor at the Graduate School of Library and Information Science at the University of Illinois at Urbana-Champaign and the director of the Center for Informatics Research in Science and Scholarship. Her research investigates the changing nature of scientific and scholarly information work in the digital information environment, with a particular focus on barriers to interdisciplinary inquiry. She has written and presented widely on information support for research communities and how to align digital research collections with scientific and scholarly practices. She leads the nationally scoped Institute of Museum and Library Services Digital Collections and Content initiative, and is co-principal investigator on the Data Conservancy, an NSF DataNet award. Her other recent funded projects include investigations of data-curation needs across sciences, scholarly annotation, and institutional repository development, as well projects to develop educational programs in data curation and biological informatics.

Arnaud Trouvé is an associate professor in the Department of Fire Protection Engineering at the University of Maryland. His interests include fire modeling, including computational fluid dynamics and zone modeling; direct numerical simulation and large-eddy simulation of turbulent reacting flows; high-performance (parallel) scientific computing; physical modeling of fire-related phenomena: buoyancy-generated turbulence, turbulent combustion, combustion-generated soot, combustion-generated carbon monoxide, radiation heat transfer, wall surface heat transfer, water-based fire suppression systems, and flash fires, fireballs, and explosions.

Charles Westbrook is a physicist at the Lawrence Livermore National Laboratory (LLNL). During his career at LLNL, he has been division leader of chemical biology and nuclear chemistry, chemistry and chemical engineering, applied physics, and computational physics. He was awarded the Horning Memorial Award of the Society of Automotive Engineers in 1991, the Thomas Midgley Award of the American Chemical Society in 1993, the Arch Colwell Award of Merit of the Society of Automotive Engineers in 2003, and the Bernard Lewis Gold Medal of the Combustion Institute in 2008. He has been a fellow of the Society of Automotive Engineers since 2008.